THE GLOBAL IMPERATIVE

Blaine Kelley

AXIAL
PUBLISHERS

THE GLOBAL IMPERATIVE

Rethinking Religion
Repurposing Science
Reimagining Society

THE PATH TO A NEW AGE
OF
JUSTICE, EQUALITY & PEACE

Blaine Kelley

LIBRARY OF CONGRESS CATALOGUING
IN PUBLICATION DATA

Kelley, Blaine
The Global Imperative. Rethinking Religion, Repurposing
Science, Reimagining Society. English.
The Global Imperative/Blaine Kelley
ISBN: 978-0-9909961-9-4
paperback

Edited by Travis Denton
Cover design and Layout by Travis Denton

Cover image: *Hubble panoramic view of Orion Nebula.*
Elements of this image furnished by NASA.
© Yury Dmitrienko

Printed and bound in the United States
by Lightning Source.

—*For Sylvia,*
My beloved and talented partner in everything.

Before we can ever know a world of fairness and happiness, we must first accept a time of discovery and change.

To be open and accepting of the changes required, and acting to bring them about, is *The Global Imperative* of our time.

Blaine Kelley

"We shall require a substantial new manner of thinking if mankind is to survive. We cannot solve our problems with the same thinking we used when we created them. No problem can be solved from the same level of consciousness that created it. Our task must be to free ourselves by widening our circle of compassion to embrace all living creatures and the whole of nature and its beauty."

Albert Einstein
1879-1955

Contents

I.
Foreward

Living Rediscovered

There I stood, a Southern white male, all dressed up, evidently educated, a fortunate and affluent guy, clearly conspicuous even in this extraordinarily diverse crowd.

The service began. Hauntingly, there came from above what must have been the voice of an angel, if one can imagine what an angel would sound like. The glorious voice came from the balcony of this historic place, the voice of a black haired, light brown skinned, young and beautiful woman, a softly lilting soprano, singing in her native Spanish. Then, in antiphonal style, a voice from down front, the deeply guttural voice of a robe-attired Jewish Cantor, singing in Hebrew. Next came the wall-shaking sound of a powerful pipe organ. For the next stanza, a nationally known Men's Glee Club joined in with their soul stirring chorus of harmonic voices.

Now, it was our turn, and there I was holding hands with people I didn't know, swaying in unison with people I had never seen before, singing the loudest I could. This old-timey Baptist Church, located in the heart of this historically black section of my home town, shook with the emotionally charged singing of the old spiritual, "We Shall Overcome," a song I had never sung before, yet filling me with something I had never felt before.

The occasion was the annual Ecumenical Service at Ebenezer Baptist Church in downtown Atlanta, an event celebrating the birthday, life and message of its most famous pastor, Dr. Martin Luther King Jr., a man who had been assassinated for his courageous work in seeking equal rights for all those whose skin just happened to be darker than mine.

On the stage for this occasion were the next President of the United States, the Governor of Georgia, the Mayor of Atlanta, the widow of Dr. King, his sister, his namesake son, and, of course, heroes of The March, Ambassador Andrew Young and Congressman John Lewis. While a few years later I would be honored to be asked by Coretta Scott King to serve on the Board of Trustees of the King Center, and later asked to serve as the Moderator of one of these annual events, this was my first time attending this annual ecumenical service at Ebenezer Church.

I must admit that my life has never been the same since that January commemoration of Dr. King's birthday. For the first time, I saw, felt, and heard what a peaceful, loving, caring, sharing, global society could be all about.

If there were ever any segregating and separating wall between this southern bred white guy and persons of a different color or culture, that wall came down that day. In its place grew a conviction that my life could be all the more enriched by my friendships and experiences with those who, in some ways, may be different from me.

What went on at that modest, but history-filled place on the occasion of that heroic saint's birthday, was, by any measure, the offering of a world changing message for all. It was an experience in togetherness, a proof of connectedness, a verification of hope, an example of love, a reconciliation with the wrong and a manifestation of spirit, an earth stirring message I came to wish all the world could receive.

This was an experience in togetherness, but it was also a living example of conviction. "We Shall Overcome" is the theme song which carried a terribly oppressed people out of oppression and segregation on to that long road to equality. If this race of deprived and abused people could hang in there,

Martin Luther King Jr. Mural at the National Historic Site in Atlanta, GA

never give up, press their way to fairness, equality and justice, champion non-violence, and celebrate their convictions with soul-filled music and a fervent plea to their God, then here was a guide for me to follow, and an example for the world to adopt.

Beginning on that day at Ebenezer, I took my first step towards a rediscovered life, the first of many more to come.

Visionary Synthesizer

I am not a noted scholar, nor a learned theologian, distinguished philosopher, preeminent sociologist, or honored scientist. My contribution and presentation is that of a "synthesizer." I am a person of long years, of hybrid genes, of broad experience, of inquisitive mind, of critical thinking and of caring spirit. Possibly of greater significance, I carry within me the skills of bringing together disparate thoughts and things and molding them into a meaningful and useful whole. I claim to possess a certain ability to look across the pieces and parts of what lies around me, and of bringing together a finished something of significance and of usefulness for all.

My parents would suggest that this strange composite of a human being was first revealed when I was a toddler on an afternoon stroll in a neighborhood park. To the constant consternation of my nanny, I was known to take two steps forward, pause, and then take one step backward, busily engaged in observing and absorbing every speck of every item which surrounded me, be it a bird singing in a tree, a flower along the path, an automobile passing by, another child on a swing, or a dog barking away. My parents borrowed a name from the title of the local newspaper, and my own children adopted it in later life. I was dubbed, "the Charlotte Observer."

My so-called "observant traits" were further stretched by my so-called "hybrid genes." I was born of a mother whose background included growing up in a small town in the South of the United States, whose forbears had been slave owning planters and served as Confederate soldiers in America's Civil War. Out of this emerged a woman whose mark was that of a compassionate school teacher and amateur artist, an individual whose mother had died in child birth, and one bequeathed to be a successor mother of unusual sensitivity, overall a loving person who was uncomfortable with change. Pair all this with a father whose background included growing up in a large city in the North, whose forbears volunteered as Yankee soldiers and whose relatives operated a station in the Underground Railway, assisting runaway slaves to reach their freedom in nearby Canada. Out of this emerged a man whose mark was that of an engineer and inventor, an individual whose first invention won Federal patenting at age 23 and whose later career was that of owning a machine manufacturing enterprise, overall a smart man, comfortable with change.

Southerner and Northerner, Confederate and Yankee, town girl and city boy, teacher and engineer, compassionate and inventive, such a mix brought me into this world enjoined to always be blending and balancing two dissonant origins.

Fast forward to my adulthood: an educated, comfortable, affluent, white, soldier, American, Christian, Protestant, husband, father, grandfather, business professional, community leader and humanitarian advocate. As I pursued my career as head of a real estate development firm developing noteworthy buildings and complexes, I also became engaged in leadership and advisory roles of probably thirty local, national and international organizations that included business, educational, cultural, humanitarian, and religious causes. Somehow, there grew within me a certain driving force. That force was an inspiration and a mandate to serve society with that which is better, to dare to cross the lines of race, gender, religion, culture, and custom, to do what I could to build buildings which offered their occupants a sense of well-being, and to serve organizations which gave society examples of moral worth.

Along the way, I served as Trustee, Director, or Advisor to organizations engaged in every imaginable endeavor: liberal arts college, university divinity schools, graduate business school, chambers of commerce, commercial banking, visual and musical arts, human rights, mainline church, humanitarian causes and real estate development. While thinking of my military service as my advance degree in everything, I served the causes and filled official roles of six different colleges and universities, from Davidson to Vanderbilt, to Duke to Harvard, to Emory to Georgia State. From developing award winning real estate projects in the southeastern United States, to leading a water and health project which ultimately served 400,000 people in the Republic of Kenya, I am fortunate to have experienced the richness and reward I have found in my participation in so many endeavors and such diverse a mix.

I must confess the important realization which has come out of all these experiences. Given that I grew up as a devoted and practicing mainline Christian, but one who could see beyond the norms of life, I have found my faith increasingly challenged by my encounters with science and society. I have experienced a revelation which suggests that the wonders of the Universe, from the very largest to the very smallest, seem to offer a more solid and factual grounding of essential moral truths than do those blindly adopted religious beliefs offered by way of the ancient supernatural.

I find that the world's clear and certain march towards the globalization of society and the increasing empowerment of the individual are, together, giving us positive advancement in certain ways, but also leaving behind continuing challenges of moral purposefulness and personal worth. What does a society gain if it is so much quantitative and so seldom qualitative?

It appears to me that we are entering a new axial age, a notion which is explained more fully in the chapters which follow. As evidence of oncoming big change, note that global society has already begun its journey towards universalization. This change is not coming in the form of an ideally revolutionary "one government" led by a much strengthened and reformulated United Nations, but instead, change is coming in a simple and slow coming together in language, looks, dress, culture, marriage, knowledge, communication, mobility, trade, laws, governance, processes and practices.

If you want proof of this beginning movement, note the interesting, almost amusing, adoption of blue jeans as the attire of choice by the people of the United States and Europe and Russia and Japan and now even China. Note the overnight adoption of smart phones as the worldwide device of communication and connection. Note the accelerating adoption of the English language as the official international language. Strangely, though, humanity seems to be experiencing so fast-paced a journey that its travelers have not yet figured out where, how and why they are going.

With this high minded hope and wish, I also am keenly mindful that there are broad ranging societal disparities which we must first set right, and there are deep-running societal beliefs which we must rethink. These are matters we must first address. Then, we must peer into the revelations of modern day "big science" discovery, and the miracles, morals and meaning which await our understanding. The overriding composite of the societal, the religious and the scientific offers ample reason for us to rethink our thinking, and to attach our lives to a new moral and social imperative.

II.
Perspective

Axial Change

In 1949, eminent German philosopher Karl Jaspers published a book entitled, *Vom Ursprung und Ziel der Geschichte*, later translated into English, *The Origin and Goal of History*. In his book, Jaspers observed that there was an extraordinary age in world history, a time worthy of noting and worthy of naming. He dubbed this special time as The Axial Age, his definition for an extraordinarily pivotal time in world history.

Philosopher Jaspers noted that, during the period 800 BCE to 200 BCE, similar revolutionary thinking blossomed in Persia, India, China, and Greece. Jaspers observed that there appeared in each of these nation-states a new development of religious and philosophical thinking of striking parallel, a happening without any direct intercommunication from one region

"Karl Jaspers", 1983, Oldenburg. Photo: Christa Baumgärtel.

to another. More specifically, Jaspers declares that during this period of history "the spiritual foundations of humanity were laid simultaneously in China, India, Persia, Judea and Greece, and these are the foundations upon which humanity still subsists today." Within this time emerged the seeds of Jainism, Ahimsa, Buddhism, Taoism, Confucianism, Zoroastrianism, Judaism, Greek philosophies, and later Christianity, and still later, Islam. Modern day theologian and author, Karen Armstrong, wrote in 2006 of these times as "The Great Transformation: The Beginning of our Religious Traditions."

Here in the 20th and 21st Centuries, we are experiencing the signs of another "Axial Age" or "Great Transformation," the scale of which is even larger and the significance of which may be even greater. I see the seeds of this new age as beginning around 1900 CE with an explosive succession of discoveries and revelations which follow, this time led by what I call "Big Science."

While Jasper's Axial Age is defined as a time of the birthing of extraordinary new philosophical and religious thinking, I see our new Axial Age as a time of the birthing of extraordinary scientific and technological discoveries, revelations leading humanity to extraordinary material, cultural and moral change—change that is now offering extraordinary moral and material advances for an emerging global society.

The propelling turning point of this new "Axial Age" will prove to be the discoveries in "Big Science," that is: astrophysics, nanophysics, particle physics and neuroscience, discoveries compelling significant rethinking of the religious and philosophical beliefs born in the First Axial Age. With this rethinking may ultimately come the revolutionary possibilities for a global society, the possibility of universal peace, justice, equality, sustenance and understanding. The consequential question for us is not whether axial-scaled changes are upon us, but whether we will turn these changes into a grand advance for the betterment of all society, or whether we will miss this opportunity for an "axial-scaled" advancement of the human condition.

I see the challenges of our time as lying in the notion of whether or not we have the courage to rethink what we have believed in the past, and whether or not we have the courage to seize upon a Big Science driven new perspective as the path to the betterment of humankind in the future. Our challenge is whether we have the courage to cast aside the limitations of our past and the hypocrisies of our present time, turning instead to the revelations of modern Big Science, and the opportunity for achieving the wonders of what can be.

What happens next is up to you and me and our global neighbors, those of us who now live in an axial time of global change and an axial time of human advancement.

III.
Rethinking Our Thinking

Axial Opportunity

The latest scientific discoveries about the incredible vastness of the Universe and the immeasurable smallness of its composition, when viewed in their totality, likely offer humanity a more profound, factual and relevant foundation for personal religious faith than do the current brands of faith proclaimed in the churches, cathedrals, synagogues, temples, shrines, mosques, schools and seminaries around the world.

Throughout the history of civilization, humanity has found it universally essential to attach itself to some form of faithful hope, meaning that there is an essential and embedded "religious" characteristic of humanity which science must recognize and include.

While accepting and building upon the acknowledged core of goodness still to be found in much of today's religious practices, this new view of faith and belief proposes to more effectively and realistically advance society on to universal peace, justice, equality, openness, caring and mutuality, all for the benefit of humankind, and for the sanctity of the Earth on which we depend.

It is possible that never before has there been so great an opportunity to begin a transformation of humanity into a universally believing, mutually supportive, positively motivated and truly global society. To that end, we must employ both the incredible advances in factual science and the beneficent goodness still extant in the origins of many religious beliefs.

Religion and Science, Then and Now

There is within me a daring and tempting presupposition that human evolution has advanced such that science and religion would do well to learn from each other. Religion must recognize that supernatural claims about life and existence are not supported by factual science. Science must recognize that it too often offers abstract technical advances devoid of addressing the human hunger for purpose, togetherness, warmth, love, caring and sharing. To suggest that science and religion need to reconsider their relationship may go against the historic disconnect between science and religion, but a new and closer look is now the imperative of our time.

We are advised to start with the earliest beginnings of civilization. Interestingly and significantly, during the earliest times of primitive society, science was the religion. Research concerning the life of humankind at the earliest signs of civilization suggests that, quite universally, religion took the form of the worship of nature. Put another way, earliest religious thought and practice focused on the human dependence on natural science, that is, the warmth of the Sun, the nurture of rain, the continuity of fertility, the hope of afterlife, and the plea against harm. Most often faith, belief and wonder were experienced through the worship of nature's human assigned gods. There was a time when science and religion were interconnected as the same.

Beginning in the third millennium Before the Common Era, evolutionary advances began to move humanity beyond the simplistic, yet profound view of life into the realm of organized living, of organized civilization. For example, observe the shift from the time of free roaming hunter-gatherers to planted agriculture and fixed settlements. Religious thought and practice moved away from the powers of the mystical beyond to the power of that which is most immediate: to human leadership. Thus emerged the all-knowing, all-powerful prophets, pharaohs, emperors, kings, czars, chiefs, and priests. When survival-based obedience to nature gave way to fear-based obedience to humans, science and religion began to part ways, and the relationship hasn't been the same since. It is evident that science and religion have come to view each other more often as foes than as friends.

Conscience-Pricked Rethinking

As to the credentials from which I speak, it is important that I share with you that, from birth, I have been a believing and practicing person of religious faith. More specifically, I could be viewed as a pedigreed, thoroughbred

Christian, Protestant and Presbyterian. My upbringing included a regimen of Sunday school and church. My education included graduation from a church-related college. My marriage began by meeting and marrying my wife in a church. My activities included roles as teacher and officer of my church. My service included the Advisory Board of two divinity schools, Vanderbilt and Harvard. My leadership included trustee and advisory roles at several nationally respected educational institutions. While making no claim to perfect piety, I suggest that there is a solid background within me of immersion in, and knowledge of, the beliefs, liturgy, creeds and practices of mainline Christianity. Concomitantly, I always have made certain that I include a knowledge of, and respect for, the beliefs and practices of the world's other major religious faiths, including those with which I may disagree.

In me is embedded a song which I was taught at the very beginning of my formative years, perhaps at four or five years old. The song: "Jesus loves me, this I know, for the Bible tells me so. Little ones to Him belong. They are weak, but He is strong. Yes, Jesus loves me. Yes, Jesus loves me. Yes, Jesus loves me, for the Bible tells me so." What a powerful and life shaping declaration with which to begin one's life! I must confess that now I see it all differently.

Along the way, I changed. Surely my liberal arts education must have something to do with it. But real change began to emerge as I experienced the conscience-challenging cause of America's civil rights movement. One could not have lived and worked in Atlanta, the headquarters of the American South, without being ceaselessly confronted with the rightness and wrongness of relating to persons in some ways different from one's self. Undoubtedly, my numerous travels to distant and foreign countries further exposed me to deep running cultural differences embedded in nations so truly foreign to mine. Clearly, the instantaneous knowledge offered by electronic communication has broadened and quickened my understanding of everything.

Surely, the recent revolutionary globalization and mobility of society has had an impact on my beliefs, convictions and motivations. But, even more surely, the compelling recent research, writings and lectures of progressive-thinking theologians and genius-endowed scientists have awakened me to new thinking about the core beliefs and understanding which constitute my very being.

More significantly, though, I believe that the very Christian-centered beliefs in justice, equality, peace and selflessness, all matters which serve as the essence of who I am, have pricked my conscience to examine more closely that which is being claimed in the Christian Bible, and that which is proclaimed from the pulpits of the Christian Church.

In this way, I have begun to discover and question what now appears to me to be the obsolescence, hypocrisy and self-serving attitudes of so much of that which has molded my life and work. Along the way, I also have come to find that such misguidance applies to all faiths practiced around the world.

In his book, *God in Search of Man*, Abraham Herschel may best express much of what I also have come to observe: "it is customary to blame secular science and anti-religious philosophy for the eclipse of religion in modern society. It would be more honest to blame religion for its own defeats. Religion declined not because it was refuted but because it became irrelevant, dull, oppressive, insipid. When one's faith is defined by creed, when worship becomes discipline, when love becomes habit; when the crisis of today is ignored because of the splendor of the past; when faith becomes an heirloom rather than a living fountain; when religion speaks only in the name of authority rather than with the voice of compassion—its message becomes meaningless."

Let me be specific: in this day of global perspective, instantaneous communication, international travel, scientific revelation, technical innovation, demographic shifts, rising democracy, interplanetary space travel and the spread of human rights, I am perplexed and saddened by the continuing insistence of Christian leadership and its devoted followers to so doggedly adhere to such ancient beliefs as: Creation in Seven Days, Adam and Eve, Original Sin, the Word of God, Father in Heaven, God's only Son, Born of a Virgin, Kingdom of Heaven, Resurrection of the Body, Ascension into Heaven, Sitting on God's Right Hand, Salvation from our Sins, Eternal Life, Paradise and Hell, Devils and Angels, Fall of Man, Creator God, In the Image of God, Forgiveness of God, and more.

Furthermore, for all too many, Jesus has preempted the role of God, and God has been relegated to the characteristics of a human. To the non-Christian, Christianity has selfishly set itself up as the superior faith which transcends all others. But, in fairness, I also must consider the claims of the other religious faiths. I am disturbed by so much blind adherence and obsolete thinking everywhere I look. It seems to me that this blindness and obsolescence has become a universal characteristic of all the popular religions of our time. It also seems that, with the arrival of globalized society, we have uncovered an underlying competitiveness among differing faiths, each seeking to prove one to be superior over the other and each inerrant in its claims. As long as any religious faith claims to be the best and the only, human beings will miss the warmth, caring and sharing of the search for human commonality and compassion, be it their claims for Judaism, Christianity, Islam, Hinduism, Buddhism, Confucianism or any other.

For example, consider that ancient faith and its rock-bound conviction of representing God's Chosen People, living in the Promised Land, with one single person as the father of the faithful, protected by God's ordering the extermination of one's enemies, and so much more, each being tenets of the world's oldest continuing faith.

Likewise, consider that faith which declares: its God as the only God, non-believers as unworthy infidels, genders to be separated and suicidal sacrifices as the highest form of fidelity, each being tenets of the world's fastest growing faith. Then, too, consider that belief in the birth decreed caste demarcation of its people and the reincarnation of their souls as central to one of the world's oldest faiths. Additionally, consider the dependence on indwelling self-serving meditation and the overriding essential of self-happiness as the leading factors in the world's most peaceful-minded faith.

More Specifically

- Consider why, in this day and time, there continues to be a faith where all the men believe that women are not worthy of praying at their wailing wall.
- Consider that faith which insists that "only Jesus saves," and that matters were fine until Adam ate the forbidden fruit.
- Consider that faith which fights to the death over which relative inherited the role of their Prophet.
- Consider that faith which insists that the female should not show any of her skin, drive a car or be seen with an unrelated male.
- Consider that faith which insists that one is born into an unchangeable segment of society, that one can be reborn as something else, and that a cow can be more sacred than a human.

- Consider that widely followed faith whose "infallible" leader champions the cause of poverty while insisting that only celibate males are good enough for its leadership.
- Consider those faiths where the men harshly enforce their claims that females should be kept in subservience and ignorance and should not receive an education.

I find neither refuge nor rationale in so much of all the religions of today. Where there may be glimpses of true goodness and wisdom, they are too often obscured by the claims of the misguided, ignorant, self-serving and hurtful. Not only is it difficult to reconcile the claims of righteousness and exclusivity of each and all of the main religious faiths, but also there are too many examples of using that faith and its "god" to support that which is patently wrong for its adherents, and truly hurtful for those who are not.

Observe still more religion-supported or religion-allowed practices, such as the believing in the superiority of one culture over another, the acceptance of human bondage and slavery, the colonizing of the primitive, the justification for the annihilation of enemies, the cleansing of the weak, the subordination and abuse of women, the dismissal of love between like sexes, and the distorted claims of rights to life.

In the larger realm, world history reflects an immeasurable number of examples of the stronger tribe or nation employing violence and war to force its culture, standards and beliefs on another, not because they are more morally worthy, but because the claimant is simply physically stronger.

When religious faiths grew in following to become large and influential, then rituality replaced rightness, practice became habit, conformity buried individuality, mythical became factual, worship became mandate, prophet became god, creeds became law and facts became irrelevant.

Heaven Revisited

While I was increasingly struggling with what seemed to me to be serious shortcomings, fallacious beliefs, self-serving claims, and inbred thinking of religious teachings and institutions, another profound experience came my way. Earlier I had experienced the life changing experience brought on by the service at Ebenezer Church. Before long another mind changing event came my way. It was quite a different kind of event, at another church, my own. As for most of its members and visitors, the church of which I was a member was the place for serious-thinking worship in the manner of the Presbyterian faith, that is, the recognition of Christ as the only Son of God, that believing in Christ is the only way to eternal life after death, and the belief that the Christian Bible offers all the guidance one needs for living a full and righteous life.

So, it was surprising to read in the weekly bulletin of my church that the following week's morning program would include a visual presentation on "Astrophysics and the Universe," a unique program to be given by a much respected church member, friend, and a holder of a Doctorate Degree in Physics. It was a program this curiosity-minded me did not want to miss.

Come Sunday morning the class turnout was disappointingly small. Maybe the topic was over the heads of some and not a priority for others. For me, though, it delivered another life-changing revelation.

HH 901/902, plumes of cold gases extending from the Carina Nebula giving birth to new stars.

With his power point visual presentation, my learned friend flashed on the screen some gorgeously colored, strangely shaped images of what existed thousands of trillions of miles away, off in the distant space where the stars reside. There were photographs of what were called "supernova," the dying of one massive star while providing components leading to a new one. There were pictures of the Milky Way, a galaxy of stars of which our Sun is just one among hundreds of billions. There was the revelation that our Universe is comprised of hundreds of billions of these galaxies; that our own Sun is still 25 trillion miles away from its nearest kindred stars; that an instant of light from some of these stars is so far away that it has traveled towards us for nearly a million years before we have been able to see it; that our Universe was born in less than a trillionth of one second; that the birth of this Universe took place 13.8 billion years ago and that the scale of this Universe is still expanding; that just like all living things, our Universe was born, is now living, and at some time will die; that while the Universe is expanding, everything in it continues to change and evolve, including all living things and including us human beings.

And so the story went, raising questions from its spellbound audience faster than the images could be shown. Here I was in my church, the God-decreed and self-declared home-place of ultimate truth, the fountain of wisdom and belief which guided our brains and sustained our lives. At least, that is what I had been taught, and that is what I had been believing.

Now came this story about the Universe, this immeasurably vast, ever-expanding, ever-changing something which existed way beyond our detecting and even way beyond our capacity to understand. Then it dawned on me: "Where in all this is God?" "Where in all this is the place or condition called Heaven?" These images are certainly nothing like the human-like God I had seen painted on the ceiling of the Sistine Chapel in Rome. Clearly, my doorway to life, existence, and God, had been flung wide open, and standing there in the cold nakedness of verifiable facts was a reasonable reason for

The Creation of Adam *by Michaelangelo. Sistine Chapel.*

me to ask: "Who, where, how and what is this entity that for thousands of years has been called 'Our God'?" And even bigger than this: Where in all our thinking and living do we humans fit in a Universe which includes more than a hundred septillion other suns and solar systems. That's more than 100,000 000,000,000,000,000,000,000 stars out there in space.

For me, the message was clear: there is an undeniable mandate that we immediately seek to learn everything we can about an existence which is far bigger, far more complex, and far more changing than our human brains can even imagine. Could this modern day revelation about existence not be more relevant and correct for us today than is that revelation about existence formed by a distant generation of humans a few thousand years ago?

Uncommon Commonality

Remembering that the main religious faiths of today had their beginning at some time during the 3,100 year period of 2,500 BCE to 600 CE, we can rightfully claim that not only is there a commonality in their history, but also there is a commonality in their practice. First, without exception, whether it is Hinduism, Buddhism, Judaism, Islam, Protestantism or Catholicism, religion has been too often the political force which led to wars, rather than the moral force which prevented them.

In all candor and fairness, we must acknowledge the wrongs and horrors of the Crusades and the Inquisition between the Christians and the Muslims, the annihilation of the Native Americans by the European-born Christians, the long warring between the Catholics and Protestants in Europe, the forced colonization by the Europeans of most of the Southern Hemisphere, the invasion by the fearsome northern Mongols against the peaceful southern Chinese, the horrors of the Holocaust committed by those of Christian background against those of Jewish heritage, the suicidal killings committed by the Islamic Sunnis and Shia against each other, and more recently, the barbaric atrocities of the "Islamic State" committed in the name of their brand of religion. There is also the horrible genocide committed by Rwanda's Hutus against their neighboring Tutsis, the massive holocaust of Pol Pot imposed on peaceful Cambodians and their neighbors, the atrocities of the Christian Serbs against the Muslim Bosnians, and the atrocities of the Islamic Sudanese against those of native belief.

While each of these travesties can be defined as the ultimate evil, there is one more that took place in societies which claimed to be Christian, Democratic, educated and advanced. Historically, it was not that long ago. And, it took place in a culture which declared in its founding documents that "all men are free." Never-mind that this society looked upon black men and women as property, inferior and not advanced enough to be equal. It pertains to my own heritage, culture, religion, and nation—the inhuman kidnapping of more than ten million peaceful native Africans, the transferring of these humans in ships under inhuman conditions defying description, the enforced enslavement and servitude in western world countries reaching from one end to the other of the western hemisphere, the breaking up of families, and a phenomenon lasting for hundreds of years. There may be no worse scourge in all history than that of only a few generations ago. That I have ancestors who supported this evil shakes me to the core of my being. Just maybe, this massive wrong of the past is one thing that has prompted me to seek an axial answer for the future.

There is often an irrefutable characteristic of dissonance and death within and among the very people who most strongly adhere to their religious beliefs, further insisting on the infallibility of their beliefs and the superiority of their culture. Too often the religious beliefs of groups of people are held together, sustained and nurtured by a man-made mandate called "dogma." In effect, the dogma and creeds of the faithful not only become the marks which distinguish them, but also, become that which is used to proclaim their superiority and to enforce their adherence.

And, since the earliest times of history, the need to believe in an afterlife, a centerpiece of nearly all religious faiths throughout history, has become the commonality among faiths and the factor of belief giving religious organizations their most effective force for sustaining their existence and for retaining their adherents.

While so often different in their details, there is an undeniable characteristic which is common to most all the faiths of the past and remains common to the faiths of today. On one hand, we find that these faiths often contain elements of universal goodness. On the other hand, that goodness is often buried under overarching elements of universal badness.

Interestingly, these commonalities of the mainline faiths constitute a compounding force for both good and for evil. Thus, to me, they suggest the essential need not to rest with the present, but to continue on with a relentless and restless search for the truth, no matter the consequences for the beliefs of the present and those of the past.

Silos and Fences

If this is to become a new Axial Age of societal advancement, there is another characteristic which stands out as an impediment. It is what I define as the "Silo Society." While there are wonderful advancements taking place in many realms of our existence, there is one which seems to be going in reverse. The more gains we make in technological devices and creature comforts, or in global knowledge and global intercourse, the more we move towards specialization and separation. We already know that there is no more segregation and separation to be found than exists in the institutions of religious faiths. Surprisingly, similar specialization is found in the fields of the most sophisticated forms of science and technology.

Not long ago I visited one of America's oldest and leading universities. My objective was to learn the latest thinking and discoveries in the areas of science and religion. Expecting to find exciting new revelations which inure to the advancement of the human condition, I instead found profound examples of continuing barriers to the advancement of the human condition.

I called on a distinguished member of the Physics Department, a learned professor in the field of Quantum Mechanics. In asking of his role in other matters to be found across the learning spectrum of an internationally leading university, my host reported that he was so busy keeping up with his assigned topic of the science of Quantum Mechanics and with his assigned mandate to publish or perish, that there was no time to visit with or to consider the findings of those of any other department. In my view, this distinguished professor operated inside a "silo," (a silo traditionally being a grain storing concrete cylinder built tall and strong to keep its contents from spilling out and mixing with the outside, and no way for the outside to impinge itself on the inside). As I saw it, this accomplished scientist lived solely for the advancement of knowledge within his tightly focused field of quantum science exploration.

I next made a point of calling on the Dean of the Divinity School, that school being one of the most highly recognized schools of religion anywhere in the world. Again, my purpose was, not only to learn of the latest thinking among theological experts, but equally, to learn of their inclusion of universal learning about all faiths and about all life. Again, came a parallel response from the Dean. Their designated specialty focused mostly on the perfection of understanding of religious history and dogma, especially among the Abrahamic religions, there being neither interest in nor time to explore their interrelationship with the many other relevant and important subjects being taught at this renowned university.

I came away from my visit with the conclusion that universities of today are, in fact, not places of collective and universal enlightenment. Instead, they are the protectorates of individual specialties, conveniently gathering all in one single place for the purpose of more efficient administration and more effective funding. It pains me to think that even these institutions of highest human learning exist only as an operating base for isolated and insulated "silos" of specialization.

As a person whose professional occupation has been that of a real estate developer, I can't help drawing a parallel with that of developing a new building. I see this specialization and separation as being like developing a magnificent building, but one which has no doors or windows or stairs, no means of stepping outside to relate to the surrounding environment, and not even a way to view anything which lies beyond. Such a closeted existence

Refugees along the Slovenia/Austria border. October 2015. Photo: Janossy Gergley.

leads to a myopic and internalized view of everything, including one's own life, and one's perspective on the world beyond.

Nations can also be thought of as silo societies, or put another way, gated communities. It is the characteristic of nations to establish their identities and to reinforce their differences by maintaining military enforced borders. There are nearly 200 independently governed nations occupying this planet. Their militarily guarded borders seemingly exist to protect them from having to mix and accept the different cultures, traditions and resources of the world which surrounds them.

No wonder there is no consensus to be found on so many crucial issues—issues which continue to thwart the well-being of humanity. No wonder so much leadership energy is devoted to coping with differences instead of advancing solutions. No wonder these issues and impasses repeatedly break out into unforgivably destructive war, rather than breaking through to desperately needed peace.

There seems to be an overriding characteristic of the human species, indeed of all living matter. It is the human essential of collectiveness and togetherness. Put another way, since the dawn of the human species, and also the dawn of living things, there is an undeniable force which causes like-minded humans and like-made living things to stick together. Survival is best served by existing together in groups, be it tribes of humans, or clusters of trees, or schools of fish, and just maybe, the galaxies of stars or congregations of like believers. History reminds us that togetherness is the best defense for survival, and aloneness is the quickest path to extinction. But, does that have to mean that we will never peacefully live and work together?

It seems that the history of civilization offers us living proof that there is an absolute inbred necessity for humans to gather around matters in which there is commonality, never mind their being right or wrong, wise or unwise, just or unjust, realistic or unrealistic. If we are to advance the notions of peace, caring, sharing, justice and compassion, there lies before us an extraordinary task: tearing down the silos of society and opening the gates of nations.

Of the world's 200 nations, there is not a single instance where there are two nations whose governance, beliefs, customs, and cultures are alike. We now live in a world where differences rule the day. Are we to accept that differences and separation are part of the natural law of evolution, or do we celebrate our differences and let them meld with another into new cultures, or are we to declare that openness and togetherness are the ultimate direction of evolution? If we are to choose the latter, then we individually and collectively must allow our faith to be made of facts and our facts to be aided by faith.

Safe Harbors and High Seas

Several years ago, I was privileged to visit with acclaimed futurist and author John Naisbitt, publisher of a book entitled *Megatrends 2000*. In our visit I asked Naisbitt of his view as to why there is a human tendency towards seemingly inexplicable and stubborn adherence to old beliefs, old practices, old traditions, indeed the "safely familiar". This adherence to the past seems to be a glue which locks in place the core beliefs of nations, religions and cultures. For example, nearly half of Americans, and probably half of the world's citizens, still do not accept the concept of "evolution" and still insist that a "heavenly paradise" awaits them following their death. Put another way, if it represents change and if it reflects uncertainty, compared to what one is born into believing, then a large share of Americans wants no part of it. And, such is all too true of the rest of the world. Naisbitt wisely explains that it is the prevailing nature of human beings to need a "safe harbor," a retreat, a place, a belief that is not caught up in all the cacophonous disturbances of modern change. Some examples of this include the difficulty of early horse and buggy owners to adapt to that new monster called "the automobile;" the difficulty of early slave owners to accept the notion of equality with their slaves; the difficulty of earlier farmers to adapt to a machine-driven society; as well as the difficulty of many world citizens to adapt from a dictatorial society to a democratic country.

Change, in any form, including that for the better, is more frightening for many than their culture, their education, and their experience will allow. Thus, there is a logical explanation for one to firmly anchor one's life in the safe harbor of the past, even if it is thousands of years old, rather than being exposed to the challenges of change, even if those changes offer that which is better.

To those who may be fearful of change, consider that changing and re-thinking do not have to lead to a void, or emptiness or to feelings of loss. Instead, change can offer the beauty and excitement of freshness and wholeness.

Consider that the history of the United States is a story of change beyond that ever known before: the rapid populating by immigration of what had been mostly wilderness, the shift from the domination by a king to the establishment of democracy, the economic dependence on slavery to the elimination of slavery, the rescuing of our allies from invading dictatorships in two world wars, the shift from an agricultural economy to an industrial economy to an informational economy.

IV.
Disparities of Existence

Complementary Complications

Given the problem of societal resistance to change, with religious faith being that most basic component of society seemingly most resistant to change, there are other components of our society which are caught up in the opposite direction, a runaway pace of change so forceful that we may rightfully wonder if such a pace of change is pushing society over a cliff. There is so much change that the consequential newest circumstances are splitting society into two inoperable and dysfunctional halves. They are the halves, or what may be better defined as the disparities, which are analyzed more fully in this chapter.

It would be incorrect and unjust to weigh in on the wrongs found in the specific beliefs and practices of various religious faiths without first recognizing the significant societal disparities among human life all over this earth. Interestingly and oddly, both the religious disparities and the societal disparities share in common one basic factor. Both are most often a consequence of one's birth and less often a condition of one's choosing. Nevertheless, each generation seems to be an inheritor of resistance to change. This being the case, we are challenged to find ways to motivate others to rethink, to reconsider.

As to the societal picture, the most graphic and comprehensive depiction of the disparities across global society may best be found in a cataloging of societal data offered annually in the profoundly reliable *Pocket World in Figures*, which is published jointly by The Economist Newspapers, Ltd. and Profile Books, Ltd., both headquartered in London.

Disparities By The Numbers

In recognizing the Disparities of our Existence it is important to first define the matter in the form of summary statistics. A review of these statistics, presented in this manner, will likely prompt us to not only *"know the facts,"* but also to *"feel the urgency,"* of that which confronts our society of today. Here are some of the harsh realities with which our new "Axial Age of Hopefulness" begins—the cruel challenges into which our newest generation is born:

WAR

Since 1900 and the great promises of the new Industrial Revolution, there have been 68 wars and conflicts during which around 200 million human beings, military and civilian, have been killed. That's nearly 5,000 humans tragically killed or dying every day since 1,900, simply because someone, somewhere could not get along with another.

POVERTY

According to the World Bank Development Indicators, in 2008 almost half the world, that is over 3 billion people lived on less than $2.50 per day,

Leh, India—June 29, 2015: Unidentified poor Indian beggar family on street in Ladakh. Children of the early ages are often brought to the begging profession.

while more than 80% of the world's population lived on less than $10.00 per day. On the other hand today, according to the World Bank Development Indicators 2016, in 2012 13% of the world's population or 910 million people lived on less than $1.90 per day. Compounding this, the poorest 40% of the world's population accounts for 5% of global income, while the richest 20% accounts for 75% of the world's income.

HEALTH

More than 1.1 billion people in developing countries have inadequate access to clean water, and more than 2.6 billion people in developing countries have no basic sanitation facilities. Most often due to lack of sufficient clean water and the absence of effective sanitation, 22,000 poverty stricken children die each day.

JOBS

Youth between ages 15 to 24, make up 17% of the world's population. What is significant is that 40% of these are without a job, and not attending school. In the Middle East and North Africa more than 25% of the young are without jobs. In some parts of Europe, specifically Spain, Portugal and Greece, more than 50% of the younger generation are unable to find employment. Even in the United States, during recent times, millions of young people have been unable to find meaningful full time employment at a livable wage.

EDUCATION

According to Buildon.org, an organization which monitors and compiles education and poverty data, around the world as of 2014, there were still 900 million adults who could not read or write and 57 million primary-aged children who were not in school. Then, too, only 6.7 % of the world had a college degree. There is a vast knowledge gap between the unfortunate masses and the fortunate few. Of the world's societal challenges this one seems to be the most disturbing because for the unfortunate, the world's knowledge base is increasing much faster than this group has the education to understand it. Ironically, while there may be advances in learning here and there, and there now may be devices which help expedite the learning process, whole societies and nations still seem doomed to fall further behind.

It is reasonable to summarize this overarching statistical story by declaring that it seems that half the world is preordained to lose, to struggle, to suffer, to strain to survive, while the other half of the world is their beneficiary. The world's disparities exist in many forms, but war, poverty, health, jobs and education are of the greatest urgency.

Disparities of Technology

Other than the explosive birth and expansion of the Universe brought on by the Big Bang event, there may be nothing else to compare with the explosive changes on Earth now taking place in the realm of technology, brought on this time by the inventiveness of the human mind.

Not very long ago, transportation meant riding on a horse; today, transportation in the air can move one along at nearly the speed of sound. Not very long ago, information was communicated by tapping a telegraph key; nowadays, messages and images are instantaneously sent by tapping a hand-held electronic device. Not long ago, it took weeks and months to deliver a message across the ocean; nowadays knowledge is instantly distributed around the globe by way of satellites orbiting in space. Not long ago, travel to a foreign country was a lifetime event; now astronauts travel to the moon in a matter of hours. Not long ago, television was sending us images of events around the world; now telescopes in space are sending us images of stars trillions of miles away. Not long ago, illnesses were dealt with by hope and a prayer; now, the marvels of medical advances save lives with miracle drugs and laparoscopic surgery. Not long ago, the destructive science of war was delivered by foot soldiers on a battlefield; now, even greater destruction is delivered by an earthbound drone pilot sitting safely in a control center thousands of miles away. Not long ago, to locate and reach a distant destination, one looked at a printed paper map; now, satellite-borne guidance systems show us the way to a precise spot at a location anywhere on Earth.

With each new generation, we seem to be increasingly becoming a world of detached and depersonalized relationships—seemingly a society more focused on the flood of mundane messages than on the pursuit of the well-being of our neighbors.

In many ways and in such a short time, a revolution in technology is pushing us into a new world. The societal implications of this change seem to be split into opposite directions. In one direction, humankind is much more informed, much more in touch, busy tearing down tribal barriers and national loyalties—all a good thing for the sake of humanity.

In the other direction, humankind is challenged to retain its humanity, to relate in person, and to find a place for quiet and reflection. In these strange new ways, humanity is experiencing the multiple benefits offered by technological inventiveness, such as in communication, travel and information, while it appears to be at the expense of lessened personal contact and direct participation.

Disparities of Demography

As technology is bringing on one kind of revolution, demography is experiencing another. In the area of demographics, there is a disturbing advance of disparity, inequity, and upheaval running parallel to the technological revolution. There is within the qualitative measure of humanity a splitting down the middle of that which is good and that which is not, as illustrated in the statistics which follow.

A review of the *World Factbook* published by the Central Intelligence Agency of the United States reveals a significant difference among the nations of the world as to their birth rates, population growth and population ages. With average birth rates often less than 1.5 children per woman and an average age of around 44, the more affluent and educated nations of Japan and Europe are experiencing a gradual decrease in population. With average birth rates of over 4.0 children per woman and average ages between 16 and 19, the nations of Africa and the Middle East are experiencing a crushing increase in their population of the less educated and less fortunate. Put another way, lesser educated and lesser affluent African, Indian and Middle Eastern societies are growing at a rate of around 3% per year, while better educated and comfortably affluent European and Japanese populations are not even replacing themselves. The significance of all this: The less fortunate are increasingly overburdening and overtaking the number of the more fortunate, and the non-productive are increasingly overburdening the productive.

In the United States, another demographic shift is taking place. Here, the share of the population over 65 was 4% in 1900, 12% in 2000, and is projected to advance to 20% by 2050.

Overall, the world population is accelerating exponentially in growth, having been one billion human beings in 1800, four billion only 150 years later, and now seven billion only 50 years after that. Relative to this, in his 1798 essay, *The Principle of Population,* Robert Thomas Malthus, demographer, scholar and minister, warned 200 years ago that the continuing growth of the world's population would exceed Earth's capacity to sustain it.

As to the composition of the growth in the United States, the historic societal domination by whites, in general, and white males, in particular, is fast diminishing. In its place there is an accelerating mixing bowl of diversity with a ten-year increase of Hispanics at 43%, of Asians at 43%, and with South Pacific people at 36%. The United States, its races, its genders, its ethnicities, its colors, its faiths and its languages are changing in a way unimaginable only a generation ago.

Fifty years ago, when a traveler waited at the Atlanta airport to catch a taxi, and then inquired of the driver about his background, the answer came from a middle-aged, southern-speaking, white male, born in a small town no more than a hundred miles away. Now, at what has become the world's busiest airport, when a taxi driver is approached with the same question, the answer is likely to come from a quite young, foreign-sounding, darker-skinned person, born in a distant land such as Nigeria, Ethiopia, Sri Lanka, India, Haiti or Nicaragua.

The mixing bowl of humanity is churning faster and faster. There are valid reasons to ask: are these changes bringing society more complications than it can process and absorb, or are these changes bringing a compensating and healthy dynamic to the well-being of society?

Along with these churning changes of demography comes an overriding moral question. How do the more privileged nations deal with an often overwhelming tide of immigrants who are understandably seeking a better life and refugees who are desperately seeking a safer life? The recent flood of refugees escaping the deadly ills of their native Syria offer us a prime example of the disparity between those challenges being experienced by the escaping desperate refugees and those challenges being experienced by the nations pressed to receive them as their new home.

Disparities of Resources

The world's two most critical resources (after air, that is) are in the process of swapping places. Once again, this is a good news – bad news scenario. For most of the world, and for most of civilized time, there has been an abundance of water, at least enough for respective nations and people to survive. The explanation for this good news is that the world's population was smaller, and part of the news is that nature has blessed most people with a convenient supply sufficient for their needs.

The sufficiency of oil and gas and coal is another matter. Once upon a time, thanks to a smaller population and a kinder nature, the need for coal for heating purposes somehow managed to balance out. Then, over the past hundred years, four big things happened: automobiles and airplanes came along, air conditioning became common, infant mortality rates improved and people began to live much longer. While the supply of water remained comparatively sufficient, the sufficiency of energy supplies became a new problem. Suddenly, the more developed nations of the world became dependent on the petroleum abundance of a limited number of lesser developed nations. Beginning with World War I, the big consuming nations of North America and Northern Europe, in order to fill their energy needs, found themselves doing big business with nations whose culture, language, governing and religion were completely foreign to that of their own. Too often, this imbalance of everything has become one of the root causes of war, violence and hate. While the imbalance between water and energy is reversing for many nations, war, violence and hate continue on.

In North America, we are now finding an abundance of natural gas and saturated shale so sufficient for our needs that the production of oil and gas is leading to an impending shortage of that other precious resource: water. There are those who understandably delight in the news that the "fracking" of oil-soaked shale in North America has brought independence and liberation from the troublesome nations of the Middle East. There are others who rightfully remind us that the "fracking" process often requires the consumption or contamination of more water than is the quantity of fuel produced.

What lies ahead is an overall sufficiency of energy and an insufficiency of water. This dynamic alone will fast bring societal change and national tensions to a degree worthy of great concern.

Disparities of Wealth

It is quite evident to all, whether they be American, Chinese, Nigerian, Afghan or Brazilian, that there is an unspeakable disparity of wealth across

New York Stock Exchange: Symbol of Amercia's Wealth

our world and even within our own country. The annual gross development product, that is, the aggregate production of goods and services, in the top ten wealthiest nations averages around $60,000 per person, while in that of the ten poorest nations the average is around $300 per person, a disparity of 200 to 1.

The United Nations recently prepared a world measurement defined as the "Human Development Index," a reflection of income plus education plus life expectancy. Not surprisingly, the spread is around 4 to 1, with the United States near the top, and again, sadly, the African nations at the bottom.

Most significantly, the greatest disparity of all lies in the United States. According to the Congressional Budget Office, between 1979 and 2007 income of the top 1% of Americans grew by 275%, while the income of the 60% composing what we call "the middle class" grew by just 40%. During the economic expansion between 2002 and 2007, the income of the top 1% grew 10 times faster than the income of the bottom 90%. In 2007, the top 20% of Americans owned 85% of the nation's wealth, while the bottom 80% owned the remaining 15%.

In his September 2013 report entitled *Striking it Richer*, Emmanuel Saez of the University of California Berkeley writes of even more recent and more disturbing news: [In the United States] "The top 1% incomes grew by 31.4 % while the bottom 99% incomes grew only by 0.4% from 2009 to 2012." The message within these statistics: the gap is growing exponentially worse.

The economic inequity among people of this world is disturbing. While for a few, the change has been positive, for most it has been negative. It is inevitable that such disparities cannot stand and such inequities cannot be sustained. Borrowing from a popular illustration claimed by those of a conservative political persuasion who suggest that "a rising tide raises all boats," I suggest that their declaration may offer a truism about boats, but the above statistics prove that such a truism simply falls apart when it comes to humans.

It is inevitable that change must come. Exactly what that change is to be may remain unclear. However, one thing is certain about the past: traditional religion has not helped, and inventive technology has been compounding the problem.

Disparities of War

War represents another radical and rapid change within and across world society. Strangely, the nature and methods of war of today appear to be returning to the same to be found in the primitive times as civilization emerged. In prehistoric times, there was the individual axe wielding person who could cause harm to a few others. Today, there is the individual, suicidal, explosives-laden terrorist who is causing harm to some. In between, there has been an evolution from masses of sword-wielding, armor-covered warriors battling to prevail, to hordes of horse-borne, spear-carrying fighters racing across fields, to mechanized tank-protected soldiers seeking to outgun the other, to massive invasions by land, sea and air intended to overwhelm the enemy, to dropping nuclear bombs for instant annihilation of masses, up to today's explosive-bearing, suicide-bent youngster committed to the spreading of terror as a devious device for delivering death to those they label "infidel."

We also must face the fact that the world's most advanced and most fortunate nation, the United States, spends more money per year on defense and war—between 500 and 700 billion dollars—than do the next 15 nations combined. These 15 nations include, collectively, our historic enemies, China, Russia, Germany, Japan, Iran, North Korea, and others. Do these numbers demonstrate that the United States is the great protector of the good, or do they tell us that the United States is a nation more seriously despised

Bashik Frontline, Kurdistan, Iraq—Two unidentified kurdish (peshmerga) fighters in back of truck at Bashik (bashik) base 25km from ISIS controlled Mosul.

than any other? More likely, these statistics tell us that there is something of great significance about the tensions, the differences, and the disparities which continue to exist among the nations and cultures of this world.

The grand irony is that, while the methods of war continually evolve, the end product of death has not. Humankind has not yet freed itself from inflicting death as the leading device for getting its way. The other grand irony is that religious beliefs and customs too often have been a major factor for causing war, rather than being the means of preventing it. One would think that by this time of so many advances in civilization that war would now be obsolete. Instead, violence committed in the name of religious faith or cultural standards seem to be increasing. Again, there is no better proof of this than the recent emergence of ISIS and their declared goal of establishing a caliphate, or Islamic Nation.

Humankind, so far, seems to have it all backwards. We have evolved into being more clever at advancing the methods of war than we have at being clever enough to rid ourselves of the reasons for war. In this time of so much inventiveness, we must ceaselessly seek ways to employ the goodness in religious faiths and the cleverness of scientific inventiveness to reverse such craziness.

Disparities of Gender

As in so many components of our existence, the views and practices of that most basic function of our humanity, sex, have been experiencing revolutionary change. Once again, some of it may be considered good, but so much of it is not.

In some nations and societies, sex has come to be the reason for segregating, covering, demeaning, or hiding the female by the male. In Western societies, gender appears to be increasingly identified by the wearing of minimum attire, such as the young female wearing of "short shorts." In other societies, gender is identified by the maximum covering of the human body, such as the female wearing of the "burka." Here lies another example of the extreme diversity across humanity. All too often, religious related views are at the root of whatever the cultural practice.

It is in the oldest and most insular forms of religious faiths that we find the most obsolete practices and thinking. Among the Hasidic Jews living in the heartland of religious beliefs, the men have declared that only males are qualified to pray at the ancient "Wailing Wall. In America, almost every coeducational college and university has been reporting a disturbing increase in incidents of rape and sexual abuse.

Muslim veiled women in the heart of downtown Istanbul

While there is good reason to wonder about the future trends of sexuality in the modern and more educated nations, there is far more reason to worry, even panic, about the mistreatment and demeaning of women and girls in such societies as the Taliban in Afghanistan, the Wahhabi in Saudi Arabia, the ISIL in Iraq, the Hasidic in Israel, and the lower castes in India, among others.

A reading of the daily news reveals an ongoing flow of cases of abuse. We learn that India's national crime records bureaus show that a woman is raped every 20 minutes. In another instance, we learn that in Pakistan a visiting

Norwegian woman was sentenced to several years in jail for "allowing" herself to be raped by a man she was dating. In another report we learn of the gang rape and killing of a woman because she wanted to marry a man of another religious faith. There is still another media report about another serious disparity, this time from Uganda, where the government announced that homosexuality is a crime to be punished with death by stoning. All too often, the practice of such inequities and cruelties are supported or allowed by cultural disparities rooted in religious beliefs still carried forward from ancient times.

In a recent meeting of the Board of Councilors of The Carter Center in Atlanta, the attending group heard former United States President and Nobel Peace Prize recipient, Jimmy Carter, declare that the wrongs of gender abuse and inequality around the world are so severe that the Carter Center, as a global serving organization, and he, personally as its leader, have now reset as their number one top priority addressing the widespread mistreatment and demeaning of women around the world. In his 2015 book, *A Call to Action,* America's former President, Jimmy Carter, declares that the world's discrimination and violence against women and girls is the most serious, pervasive, and ignored violation of basic human rights.

As in other matters, surely such societal wrongs will not survive in the glaring light of global exposure. But what will it take to jar the reasonableness in society to bring change and sanity to an already burdened world? Will it be something from religion? Will President Carter's work suffice? Is better education sufficient? There is evidence that the matter is going to require all these, but much, much more.

Disparities of Race

If today you were to ask a typical white American citizen, "Do you feel that America has now arrived at true racial equality and justice?" The answer, most likely, would be an unequivocal "Yes." Ask that same question of a black American, and most likely the answer comes back with an emphatic "No."

Many in our nation, and so many around the world, applaud the fact that America has set a

Dr. Martin Luther King, Jr. speaking at the Civil Rights March on Washington, D.C., August 28, 1963.

new world standard: It elected a black man as the most important leader in our nation, our President. While the positive feelings and pride may not be totally universal, no one can argue that a significant change has taken place in America.

Before those of white heritage get too comfortable with their notion that the high noon of social equality has finally arrived, they need to listen more closely to the words spoken at the commemoration of the 50th anniversary of the historic March on Washington, and they must recall the words of its leader, Martin Luther King, Jr., who gave to the world his never-to-be-forgotten "I Have a Dream" speech from the steps of the Lincoln Memorial in Washington. On that 50th anniversary, speaker after speaker reminded the world that disparities remain and that true justice and equality continue to elude us.

Those messages could have been given from the plaza of the Kremlin, or the dirt crossroads of Kenya, or the Christos hilltop of Rio Janeiro, or the Revolution Monument of Mexico City, or the Tiananmen plaza of Beijing. While advances are being made here and there, the vision of the mountain-top dream of Martin Luther King has not yet been seen by his nation, nor by that of any other.

If there is any doubt that something is not yet right, just visit the bastions of righteousness, the Christian Church or the Jewish Temple. There is still no more segregated a scene than in the churches of the southeastern United States. In fairness, while historically "white" churches of the South are purist in the color tone of their congregations, so also are the historically "black" churches.

But let's not stop there. Go visit an American Korean church, or an Amish meeting house, or a Greek Orthodox cathedral, or a Hindu Temple in Nairobi, or a Shia Mosque in Yemen, or a Buddhist temple in Laos.

These are proof of something which says that in the world's most sacred sanctuaries of "brotherly love," there is a more pure racial, gender and cultural divide than may be found anywhere else. If we want to fling open the doors to the greater good, we would best start with the rethinking of that which we hold most sacred, our institutions of belief and worship.

Disparities of Knowledge

Clearly, one of the most astounding phenomena taking place on this planet is the growth of knowledge. As reported by Norman W. Edmond at the 1992 conference, "Teacher Education for the 21st Century," information and knowledge are expanding at a rate far beyond our imagination. Dr. Edmonds reports that "From 1750 until 1900, it took 150 years for the amount of knowledge to double. Knowledge doubled again in the 50 years between 1900 and 1950. It doubled again in the ten short years between 1950 and 1960. Currently, knowledge and information are now doubling every three to five years. By the year 2020, it is expected that the quantity of knowledge and information will double every 73 days." Possibly no other component of our society is expanding at such an exponential rate. The big questions are: Can the human mind keep up with this runaway expansion of knowledge? Can educational institutions keep up with this runaway burden of educating? Most important of all, humankind is caught in an increasing exponential gap between those societies and those persons who are fortunate enough to be better educated and informed and those societies and persons who are unfortunate enough to be left behind.

For evidence of the impact of educational differences in the United States, the U.S. Bureau of Labor Statistics in Washington reports for the year 2014 on two key measures of population well-being. First, less than 3% of those who have earned a college bachelor's degree or higher are unemployed, while more than 9% of those who have not completed a high school education are unemployed. Second, those who have earned a bachelor's degree are receiving average weekly earnings of $1,101, while those who have less than a high school diploma are receiving average weekly earnings of $488.

Disparities of Religion

There is no single segment of all society about which more has been said, more written, more spoken, more sung, more celebrated and more worshiped than has been the subject of religion. There also is no single component of our existence about which more has been debated, fought, killed, divided, changed, sacrificed, disagreed and believed. It is perhaps the strangest and saddest part of existence. It is so because religious belief has become the model and pinnacle of disparity, when it is meant to be the ultimate guide for the conduct of our lives and the standard for our existence. Behind every form of religious faith of today there lies a history of dissonance and division.

As for the existence of God, there are extreme differences between the non-believing of today's most preeminent scientists like the late Stephen Hawking and Richard Dawkins and that of the succession of infallibility-claiming Catholic Popes, beginning with St. Peter in 32 CE and extending to Pope Francis of today.

For the Christian: Sharply differing theological claims about the divinity versus the humanity of Christ emerged in the first few hundred years following his death. These opposing differences about the nature of Christ were represented by two priests, one named Athanasius who insisted that Christ was selected by God as a divine being, and not human. The other, named Arius, who insisted that Christ was simply an extraordinary human, and not divine. This basic disparity was settled when Emperor Constantine convened the First Council of Nicaea in 325 C.E., a gathering of three-hundred Bishops who declared following a wide majority vote that Christ was indeed divine,

People praying in a Caodai temple in Vietnam. Caodai is a Vietnamese religion mixing different religions from around the world.

and not just human. Nevertheless, such differing beliefs about the life and death of Jesus Christ continue to this day, and can be observed to exist among numerous passages of the present day Christian Bible. These marked differences in belief and conviction continue on today between the literal believing fundamentalist and the broader believing progressive.

For the Hindu: The earliest forms of Hinduism emerged thousands of years before the time of Christ. The various forms of Hindu of today are the derivative beliefs and practices of successive inhabitants of the Indus Valley, the melding of the earlier inhabitants and invaders, the Indus, the Aryans and the Dravidians. There is today no single or pure form of Hinduism. However, there are three surviving general beliefs and cultures, Shaivism, Vaishnavism and Smartism, each representing a marked religious belief, differing one from the other.

For the Jew: Their very origin as a faith is described in the Hebrew Book of Genesis. There we learn the early story of conflict between two sons, Jacob and Esau. Today, we find the Jewish faith to be further divided into the Orthodox, the Conservative, the Reform, the Reconstructionist and the Humanist.

For the Muslim: The basic difference and split within Islam is big and is serious. There are two basic factions, the Shia and the Sunni. This split began with the serious disagreement around 1500 years ago as to who should inherit the sacred leadership role of the deceased Mohammed. That disagreement continues to this day, only it has become a deadly difference which today marks the ongoing split among the dozen nations which comprise what we call The Middle East, a major part of this Earth more conspicuously marked by its differences and infighting than by its commonalities and contributions.

There exists on this Earth more religion-related sects, more subsets, more denominations than this dissonance-burdened Earth can count, much less sustain. The ultimate question appears to be: If faith is the device which sustains the human through difficulties, death and differences, then why do its excesses and distortions, instead, seem to dominate the beliefs of its followers? More importantly, how can we break this destructive lock on humanity? What can we turn to in order to break us out of this seemingly death grip of religion?

At the time of the writing of this book, the possibility of a war of worldwide dimensions is taking shape, with the Islamic Sunni followers lining up under the leadership of Saudi Arabia and its allies, the Islamic Shia followers lining up under the leadership of Iran, with the ISISTs and Jahadists somewhere in between. Such violence appears destined to tear apart the entire Middle East. This, in turn, carries with it implications for a crisis in the world's oil supply, leading to a possible crisis in nuclear military strikes. These clashes and tensions remind us how two World Wars originated in the not too distant past. They also remind us of the long ago succession of wars between the Muslim Ottomans and the Christian Crusaders. With all the world's advances in globalized society, democratic governance, open communication, individual empowerment and energy abundance, there are still such deep rooted cultural jealousies, religious differences, national pride and entrenched ideologies that the people of this world have not yet been able to live with themselves, and we have begun to wonder if or when they ever will.

Reasons To Rush

In laying before us the multiple disparities of our age, I take no pleasure in being a conveyor of doom or of waking the sleeping. I hope that I am seen, not so much as an alarmist, but as a determinist, one like those good people at Ebenezer Church who proclaimed "We Shall Overcome." In the meantime, if indeed we are to overcome, we would do well to assess that which needs overcoming. I find that the disparities of our time reveal disturbing factors which justify our reassigning priorities, justifies our rethinking, and proves our need to get going fast.

After reviewing the nine big disparities confronting our global society of today, we find that five of these big, bad monsters, Technology, Demography, Resources, Wealth and Knowledge are expanding at an exponential rate. The other four of them, War, Gender, Race and Religion, seem to be locked in an unchanging grip on all humanity. Collectively, and presently, the big, bad disparities of our present day existence are, so far, growing faster than are our will and our work to reverse them.

Let's look at all this from another perspective. In this New Axial Age, technological inventions, scientific discoveries, the vitality of competition, the instantaneous spread of information, the better understanding of others, the intermixing of cultures, the empowerment of the individual, and more, are, in a collective way, giving us the tools to more fully and quickly bring about positive change. The first step, but not the only step, is for our religious faith to catch up to a thinking in terms of today, rather than so much relying on thoughts and declarations and stories of the ancient past.

If we are to hope for real advances in the human condition, we need to rush, to hurry up. We will do well to employ the incredible technological and scientific advances now at our disposal. Real advances will come only with our adoption of a new universal and common belief, a belief which speaks to the life we wish to advance, and a belief which is rooted in factual goodness, knowledge and wisdom of these times, and not on beliefs rooted in the stories and convictions of times long ago passed.

V.
Enigmas of Existence

Two Approaches

In 1875, the already famous English Biologist and Anthropologist, Charles Darwin, wrote his second book, this time an encyclopedic assemblage of information, entitled *The Descent of Man*. In this detailed study Darwin focuses on the descent of man from some lower form, going on to extend his findings to the powers, mind, morality, genealogy, race and sexuality of humankind. As if that were not enough, Darwin also analyzes the characteristics of insects, fishes, reptiles and birds. What I find to be noteworthy is Darwin's choice of the word "descent." I suppose such thinking comes from the theological and societal perspective of his time, implying that whatever exists is an action being handed down from a higher being, an empirical definition of evolution.

Nearly a hundred years later, in 1973, another famous British scientist, Dr. Jacob Bronowski produced a book and thirteen part television show entitled *The Ascent of Man*, an explanation of humankind's ascent upward from baser forms of life. I was privileged to watch every television program of Bronowski's fascinating explanation of humankind's ascent from baser forms of life.

Interestingly, Dr. Bronowski generated his findings, not so much out of the first hand field examination of Charles Darwin, but by drawing on the work of other already acknowledged scientists, from Pythagorus, Euclid and Ptolemy, to Copernicus, Galileo and Newton, to Einstein, Darwin and Pasteur, to Mendeleev, Thompson, Bohr and Fermi, and still others who followed. Whatever the source and whatever the finding, Bronowski's story declared that life and humanity ascended upwards through the deliberate discoveries and theories of the brightest, rather than descending through the accidental and incidental habits of the lowest.

I draw this analysis, not so much for its confusing perspective of descending versus ascending, but for its lead into the enigmatic characteristics of the highest form of life, the human being.

The Ultimate Enigma

For certain, our human lives are loaded with and surrounded by an immeasurable succession of unanswerable puzzles, perplexities, questions, doubts, mysteries and uncertainties about our existence. However, there is one particular category which commands our attention. It is the propensity of humanity to biologically and culturally advance, in spite of itself. Put another way, whatever advancement we may perceive has come along, not in a simplistic uninterrupted straight line as we might expect from an omnipotent and omniscient God, but in a lurching and ratcheting path of pluses and minuses which more likely comes from the unpredictable and uncontrollable workings of evolution. As illustrated in the following comparison, it is the net gain over time which we recognize as the positive and progress of our existence.

"Violence Which Never Ends"

Prolific author and lecturer, eminent theologian and historian, Karen Armstrong, published a book in 2014 entitled *Fields of Blood: Religion and the History of Violence*. While acknowledging that her motive for writing the book is to demonstrate her belief that war and violence are rarely the consequence of religious belief, and are far more often the outcome of cultural factors, this book seems to be more often a defense of religion than a condemnation of violence. Nevertheless, intended or not, Armstrong provides us with an extraordinary commentary on the history and motivations of violence across world history and across world society. It paints a picture, not of the excusable frailties of humanity, but of the ignorant and blind and selfish continuation of human beings to employ death dealing force as the device for winning its way.

Beginning with our hunter-gatherer origins, Armstrong traces the history of war and violence all the way up to today's horror inflicting ISIS or Islamic

State. It is interesting and disturbing to observe humanity's never ending natural proclivity towards violence, with "warfare being indispensable to any premodern economy." Armstrong goes on to say that, "No state can survive without its soldiers. And once states grew and warfare had become a fact of human life, an even greater force—the military might of empire—often seemed the only way to keep the peace."

Ruins of Nagasaki, Japan, after atomic bombing of August 9, 1945.

Spanning across a study of the farmers and herdsmen of prehistoric humanity to the cultures of India, to the warriors of China, to the dilemma of the Hebrews, to the acclaimed miracles of Jesus, to the tragedies of the Byzantium Empire, on to the Muslim dilemma, and to the Crusades and Jihad,

Armstrong concludes with chapters on Holy Terror and Global Jihad. In all, Armstrong presents a disturbing but compelling commentary on the violent nature of humanity.

The reader comes away from her book feeling resigned to the inevitability and permanence of war and violence. I can't forget an experience I had with a one armed former Nazi soldier I met on Christmas Eve, with his conviction that war and violence are both inevitable and necessary, a fulfillment of evolution's notion of the survival of the fittest. But, why must this be? Is there no alternative? Is there no offsetting reward? Is there not room for any advancement of the human condition? Dr. Michael Shermer, author of *The Moral Arc* declares that all is not lost, all is not in vain. There is more.

Progress Which Continues On

American author, publisher and sociologist Dr. Michael Shermer offers a profound surprise in his 2015 book, *The Moral Arc, How Science and Reason Lead Humanity toward Truth, Justice and Freedom*. He offers convincing proof that society is indeed advancing and improving, in spite of notions to the contrary.

Dr. Shermer's research yields some sense of relief. Like the sweep of a great "moral arc," society seems to advance, justice seems to prevail, and human worth seems to increase, all in spite of the selfishness of people and the hypocrisies of our beliefs. Documented with statistics, charts and specifics, the Moral Arc declares that progress does exist and continues.

Dr. Shermer declares that there are specific gains and they are accelerating. More fully documented in his book, here is his summary list:

- Homicides are Declining
- Global Nuclear Stockpiles are Declining
- Violent Campaigns for Political Change are Declining
- Percentages of Deaths in Warfare are Declining
- Incidents of Epidemics and Plagues are Declining
- International Organizations and Relationships are Increasing
- Economic Progress of Democracies Increasingly Outperform Dictatorships
- Societal Health of Nations is Advancing
- Number of Incarcerations is Decreasing
- Political Freedom Around the World is Increasing
- Adoption of Official Abolition of Slavery is Increasing
- Adoption of Women's Legal Right to Vote is Increasing
- Percentage of America's 25-32 Year Olds with a College Degree is Increasing
- Hourly Earnings of America's Workers are Advancing
- Earnings Gap Between America's Women and Men is Increasingly Closing
- Incidents of Rape and Sexual Assault of Women in America is Declining
- Use of Contraceptives is Increasing and Incidents of Abortion are Declining
- Acceptance of Gay Rights is Increasing, Incidents of Homophobia Declining

- Legal Abolishment of Torture by Nations is Increasing
- Use of Capital Punishment is Declining
- Death Sentences for Criminals is Declining
- Incidents of the Empowerment of the Individual are Increasing
- Gross Development Product of Nations is Increasing
- Global Poverty Levels are Declining

The Moral Arc insists that organized religion and religious beliefs have played little, if any, role in the advances listed above. His explanation: these advances and improvements are more often attributed to the world's increases in education and to the global proliferation of communication.

Reasons to Try

In recognizing the continuing violence, wrongs, injustices and ills of our global society, we have every reason to ask why? But in recognizing the continuing advances, progress, changes and enlightenment of our global society, we also have good reason to continue to try. Whether one attributes societal progress to the influence of religious beliefs, or to the ongoing tide of evolution or to the spreading of education, or to the explosion of communication, or to the power of human will, something is happening, and it is good. The tug of war between violence and patience is being won, ever so slowly and ever so frustratingly, by determination. Therein lies the irrefutable reason to try, or to borrow a popular expression, "to keep on keeping on."

In the meantime, back to the "Ultimate Enigma." There is good reason for us to still ask, if the well-being of the human is advancing, as Dr. Shermer demonstrates, then why does the propensity of violence continue, as Dr. Armstrong reports?

VI.

LAWS OF EXISTENCE

Sir Isaac's Third Law

In 1686, one of history's most famous and prolific scientists, Sir Isaac Newton, published a groundbreaking treatise entitled *Principia Mathematica Philosophiae Naturalis.* This treatise reported on Newton's discovery and declaration of The Three Laws of Motion, all offering the world an understanding of the function of forces and gravity, two of the most basic laws of all existence.

Interestingly, the Third Law of Newton's three principles declares that "Whenever one object exerts a force on a second object, the second exerts an equal and opposite force on the first." Surely, it never occurred to Newton that his Third Law could also serve as an explanation of the mysterious functioning of world society. That is, how and why the collective actions of humanity seem to manifest themselves in a forward and backward, or progression and regressive pattern of behavior. Newton's Third Law can be aptly applied

to explaining the pattern of the advances and then the pushback of society's politics, governments, leadership and conduct, not only in recent times, but throughout history. As a generalization, this law can certainly be applied to the constant duel, or push and tug between the liberal policies and the countervailing conservative policies, a pattern characterizing the politics and policies of so many nations around the world.

Sir Isaac Newton, 1689.

A World in Chaos

In earlier chapters, I propose that humanity is entering a "New Axial Age of Goodness and Fairness." But, how can that be? It seemed as though the horrifying image of hijacked planes crashing into the two iconic towers of the World Trade Center turning steel into dust, all in the midst of one of the world's most important cities, undid the peace and confidence of the world's most secure and promising nation. Perhaps this earth shaking event in 2001 was the historic marking of a pushback by some against the advances of others.

Beginning in the year 2010 onward, the world seemed to convulse in spasms of protectionism and nationalism, clearly a reaction to the emergence of globalization. What had been seen by some as positive signs of human betterment suddenly were seen by others as signs of failure and weakness. The undoing of "better laid plans" seems to have become rampant and contagious. Nation after nation seemed to be experiencing examples of reversing what had recently been deemed as nations working together towards universal peace and prosperity. The historically unifying and peace-making European Union saw one of its leading members, Great Britain, vote to leave the Union and go a separate way. The world's most advanced nation, the United States, in electing Donald Trump essentially voted to set aside its worldwide friendships and treaties and to shift to an "America First" policy. This change in political direction can be characterized as a reaction against governmental policy and leadership over the last decades. In addition, as worldwide nuclear disarmament seemed to be making progress among the more dominant nations, "underdog" North Korea, and Iran threatened world survival with their newly developed and still evolving nuclear weapons. The notion that the world was slowly becoming more peaceful seemed to give way to extremist acts of terrorism by a barbaric group calling itself The Islamic State.

Perhaps the most dramatic and earth shaping change has been the move from global scaled warfare to individual suicidal terrorism, or closer to home, the "lone wolf gunman" here in the United States. While the massive scale of groups of nations fighting against groups of nations has begun to subside, the frequency of individualized terrorism has taken its place. More and more of the world's cities have experienced what one or a few individuals can do to paralyze a whole city with fear. Think of the terrorist attacks in Paris in 2015 which killed more than 120 individuals who were out for the evening to hear music or have dinner with friends. There have been additional deadly attacks in London, Barcelona, New York, Boston, Charlottesville, Las Vegas, Mogadishu and Cairo. And, we wonder who will be next.

The warring issues of Syria, Sudan and Somalia and the economic woes of Mexico and Central America have brought on a tide of millions of immigrants whose arrival in other nations has caused political and cultural turmoil in those recipient nations. Then, as if the world needed more differing national interests, the Catalonian people of Spain and the Kurdish people of Iraq have begun to push for secession and the establishment of their own independent nations.

Coming out of an historic world-wide recession in 2007, economic data shows that the wealthy became wealthier while the middle-class worker lost jobs and income. While formerly secure Western nations experienced the loss of jobs and industry, the emerging economic powerhouse of China used its financial prosperity to become the world's chief sponsor of foreign infrastructure projects. The good news that women were chosen as the recently elected leaders of Brazil, South Korea and Myanmar has given way to news of a tide of leadership corruption in those same nations. Most significantly, in America the historic and progressive election of a person of color, Barack Obama, as its President is seen around the world as having given way to the historic election of President Donald Trump who is seen by many as a dangerous step backward, and responsible for tearing apart the United States' reputation as a world leader, as well as disrupting long-established partnerships around the globe, only to name a few.

As if massive death by bombs and bullets were not enough, we are now witness to the arrival of a new phenomenon of world danger and destruction: "cyberwar." While Russia is charged with hacking into America's presidential election, North Korea acknowledges that it has hacked into the secrets of

South Korea's military. In the business world we find that an unknown source has hacked Equifax, a major data company in the United States, and stolen the confidential information of more than a hundred million customers in the United States. There have also been damaging attacks recently of the National Security Administration, which have led to the sharing of U.S. secrets and cyber tools.

It is conceivable that nations, and even rogue individuals, are now capable of shutting down a nation's electrical system, water system, financial system, missile system, and more. If this new phenomenon is coming along in addition to nuclear bombs and elusive terror, then how can we possibly be thinking positively about the future? How can we even imagine the emergence of a New Axial Age of peace, justice, caring, and equality?

A Matter of Perspective

For those who would rightfully despair of the chaos and reversal of current matters of the world, we all would do well to recall how much worse matters have been in the past. To prove this point, there is probably no more accurate measure of the well-being or the ill-being of humanity than the loss of lives attributed to war and starvation.

According to the *List of Wars and Anthropogenic Disasters by Death Toll*, published by Wikipedia, and based on more than five hundred references, over the past 500 years there have been twenty-eight wars and human made disasters in each of which over a million people have died. More specifically, five of these events account for more than 25 million deaths, while another seven events brought more than 5 million deaths.

Sadly, these war-related statistics do not account for the additionally shocking massive genocidal, ethnic cleansing and societal upheaval events of which humanity is also guilty. Author Stephen Kotkin, professor of history and international affairs at Princeton University and senior fellow at Stanford's Hoover Institute, has provided us with a compelling accounting in an article entitled "The Communist Century," published as a feature article in the November 2, 2017 edition of the "Wall Street Journal."

In Kotkin's article we learn of these damning statistics: In the years around 1932, the forced collectivization under the Stalin regime led to the loss of between 5 to 7 million Soviet lives, while at the same time, under the genocidal tactics of Stalin, more than 10 million citizens of The Ukraine perished by forced starvation. In the years around 1942 Nazi repression and wars of ra-

The battle of Stalingrad saw nearly two million casualties.

cial extermination killed at least 40 million people. Around 1960, Chairman Mao's forced program of collectivization resulted in one of history's deadliest famines, claiming between 16 and 32 million Chinese lives. Around 1975, the inhuman rule of Pol Pot in Cambodia led to the death by starvation, disease and execution of an estimated 2 million Cambodians. Other accountings tell us that the Nazi repression and wars of racial extermination killed at least 13 million Soviet Slavs and more than 7 million European Jews.

Consolation Found

Fortunately, there is another, more recent, measure of humanity's behavior which is encouraging and consoling. In a report "Globally, Deaths from War and Murder Are in Decline: The World is Getting Safer, Even If It Doesn't Necessarily Feel Like It" by writer and editor Colin Schultz of Toronto, Canada, written in 2014 for *Smart News,* a publication of *The Smithsonian Institution,* Schultz notes that according to a new report from the *Human Security Report Project,* deaths from war […] have been generally in decline since the end of World War II. […] During 2012—the most recent year for which there are data—the number of conflicts being waged around the world dropped sharply, from thirty-seven to thirty-two. High intensity conflicts have declined by more than half since the end of the Cold War, while terrorism, genocide and homicide numbers are also down." Schultz adds that "the number of war deaths has also plummeted. In the 1950s, there were almost 250 deaths per million people caused by war. Now there are less than 10 per million. 'There have been some pretty extraordinary changes and they haven't been recognized.'"

Nicholas Kristov offers additional encouragement in his New York Times article "Why 2017 Was The Best Year In Human History." Kristov writes that

"Given the rising risk of nuclear war with Korea, the paralysis in the United State Congress, warfare in Yemen and Syria, atrocities in Myanmar, and a president who may be going cuckoo, you might think that 2017 was the worst year ever." But drawing on the research of Max Roser, an Oxford University economist, and on Steven Pinker, a Harvard University professor and author, Kristov reports: "Instead, a smaller share of the world's people were hungry, impoverished or illiterate than at any time before. A smaller proportion of children died than ever before. Every day, the number of people living in extreme poverty (earning less than $2 per day) goes down by 217,000, and every day 325,000 more people gain access to electricity, and every day 300,000 more gain access to clean water [...]."

Kristov goes on to remind us "that as recently as the 1950's, the United States had segregation of races, a polio epidemic, bans on interracial marriage, bans on gay sex and bans on birth control. Most of the world lived under dictatorships, two thirds of parents had a child die before age 5, and it was a time of nuclear standoffs, of pea soup fog, of frequent wars, of stifling limits on women and, in China, the worst famine in history."

If one needs more convincing that society is advancing, even if in lurches of progress and regression, then Harvard Professor Steven Pinker provides powerful statistical proof in his 2018 book entitled *Enlightenment Now.* Professor Pinker provides 78 illustrations covering 16 categories which range from life expectancy to battle deaths, from poverty to income, from education to happiness, all of which offer us compelling evidence of the evolutionary progress of the human species.

Professor Pinker proclaims that these categorical gains are leading the human into a greater capacity to reason what ought to be, and a greater capacity to see natural science as offering a guide for our living. From this he points out that more and more the human is moving towards a faith based more on the centrality of the human and away from faiths based on the God fearing supernatural, all making for a most brilliant interpretation of what is happening to the human species of today and its hopeful expectations for tomorrow.

Back to Basics

While we learn of the never ending badness of humanity and the overriding goodness of humanity, there is still good reason to ask why world society is currently experiencing such a reversal of what has until recently been thought of as progress, that is, advances with human rights, peace, civility, equality and democracy. Why is it, and what is it about society that seems to bring us progress, but always at the price of reversal? That familiar illustration of "two steps forward, one step back" seems to be an inerrant "law of our human existence." Certainly these forward and backward dynamics bring into question our claim that we are entering a Third Axial Age, with this new age proposed as a time of real hope, peace and justice.

I suggest that, for an explanation, we go back to Sir Isaac Newton. If he were alive today, Sir Isaac likely would suggest that life just isn't a straight line of forward progression. Life is a continuous motion of forwards and backwards. Put in the modern vernacular, "life just doesn't give us a free ticket to Nirvana." And, I would add, such also can be said of that path to a new age of justice, equality and peace.

Reasons to Remember

If we are to believe that life is slowly getting better, and if we are to believe that the forward and backward motions of societal behavior are still netting out to overriding progress, then there are two reasons for us to hold to those beliefs. First, remember that "we are creatures of hope," hope for the better, no matter the obstacles. Secondly, remember that there are historical facts which prove that the past has been worse, that evolution is slowly but surely carrying humanity more forward than backward.

VII.
Realities of Existence

The Hopeful Human

While still mindful of all the disparities, injustices, wrongs, and even the evils and sufferings of society, there remains in the human being a propensity towards hope. Perhaps this is a product of evolution. Or is this a gift of a possible God? Maybe hope could be an essential for the survival of the human species. In any case, hope appears to be a living force which dwells in each of us. In fact, it is fair to ask, where would all humanity be without carrying at the core of its being a measure of hope for the better?

It is reasonable for us to think that when the notion of death enters the picture, it is not as the antithesis of hope, but it is a duality, a companion. Is not the specter of death and its uncertainties a reason for hope for something else? No matter what the odds or the issues, there is an innate and overriding propensity for humans to hope.

It is my thinking that hope is the foundation for belief. Hope is the motivating force for seeking thoughts and actions in which to believe. Put another way, beliefs are not facts. Beliefs are hopes in what may be, what could be, what is wished to be. However, there is likely to be some wishful thinking, some unreal thinking, among our hopes and beliefs. Just because we hope for something doesn't mean that it will happen, and, likewise, just because we believe something doesn't mean that it is a fact. Too much of personal religious belief and too much of organized religions are based more on hope and wish than they are on fact and reality. More specifically, across the spectrum of religious faiths, all too much of their basic substance is founded on hoping and wishing that something is true than it is on that which is knowable and provable as factually true. With that in mind, let's take a closer look at some of our religious beliefs which are based on unfounded claims. Keep in mind there is also a portion of religious faith and belief that is positive, factual and real, but first, the unreal.

Realities Beyond Beliefs

A little more than half of the world's population acknowledges that it believes in a God and considers itself as part of one of the Abrahamic faiths of Judaism, Christianity and Islam. While that leaves a very large measure of the world's population believing otherwise, there is reason to believe that participants in these three faiths appear to represent the most active and influential factor in current world affairs. Of all the world's faiths, and of all the world's population, be they believers or not, the Christian faith appears to be the most dominant force in world affairs, whether or not such is considered good or bad.

The foundation of each of these three Abrahamic faiths is a belief in a single God, a belief declared to have been decreed as an essential of life with damning consequences if rejected or ignored. For the followers of the Jewish faith, it is the Torah which sets forth its tenets and conditions of belief. For Christians, it is the Bible which sets out the history and the hopefulness of belief. For the Muslim, it is the Quran which provides the stringent guide for its beliefs. While these three religious faiths together are believed to dominate the world, they also have much in common. Each insists that there is only one God. Each believes that their one God has actually spoken to them and has laid out rules for their adherence. Each of these faiths was born more than one thousand five hundred years ago. Each faith and its official book of guidance is reported to have been written or dictated by a person or persons who were illiterate. Each faith claims that the tenets of their faith have been delivered as "the inerrant Word of God." Interestingly, much of the information reported in their respective official books includes stories and events which also are found in other non-related faiths and writings of previous times.

All of this is to say that there are numerous instances of information, guidance and mandates which seem to be the imagined beliefs of writers who actually lived at a much later time. In other words, the creeds, commandments and vows which define each of these faiths, and their multitude of subset denominations, hold a multiplicity of examples of hopes and beliefs which are not supportable by reason and fact.

Rethinking Popular Claims of Faith

• Is it realistic for us to believe in a God who presides over humans like a king, while hiding in some elusive, comfortable and remote place many call heaven?

• Is it realistic for us to believe that this God created the Universe and all in it, then began functioning like a human being?

• Is it realistic for us to believe that this unseen God gives out harshly demanding orders and commandments to worshiping subjects?

• Is it realistic for us to believe that if you don't worship this God you will be condemned to a punishing afterlife of eternal hell?

• Is it realistic for us to believe that God is a being who favors certain people while dismissing others as unworthy?

• Is it realistic for us to believe, as the Protestant Catechism proclaims, that "the chief end of man is to worship God," a perceived being whom we cannot see, feel or hear?

• Is it realistic for us to believe that when suffering strikes, it is God's punishment for misbehaving or misbelieving humans?

• Is it realistic for us to believe that there is a "Devil" enticing humans away from what is right and best for those humans?

• Is it realistic for us to believe that the only correct guide for humans living today is found in the beliefs, perspectives and writings of those authors living 1,500 to 3,000 years ago?

• Is it realistic for us to believe that this great and wonderful God would choose one particular human, a male, as an "Only Son," grant him the same powers as "God," and assign him to reign over heaven and earth forever?

• Is it realistic for us to believe that the earth was designed by a God who then declared it perfect and not to be changed?

• Is it realistic for us to believe that there was an Adam and Eve who lived 6,000 years ago and who messed up a perfect world by eating a forbidden apple?

• Is it realistic for us to believe that one chosen human bodily ascended up into God's heaven to sit at God's right hand and to live forever after as "Lord of All?"

Reasons to Care

Although we find misguided claims, false infallibility, wishful thinking, self-serving declarations, limitations of history, human errors and a litany of other excuses for incorrect statements and stories reflected in the "official" books of these well intended faiths, we likewise will do well to also consider the goodness at their core. Each of these faiths attempts to offer a moral compass to its people, especially keeping in mind that these beliefs were born in a time of extraordinary abuse, loss, cruelty, hopelessness and inhumanity. In spite of the ills and adversities which surrounded the birth of each of these three faiths, it appears that the Jewish faith deserves credit for giving birth to needed ideas of order and survival, the Christian faith deserves credit for giving birth to the needed ideas of love and caring and forgiveness, and the Islamic faith deserves credit for giving birth to needed ideas of humility and unity.

While acknowledging the shortcomings and wrongdoings of each, we would do well to give credence to those faith-related contributions which we find measurable, relevant, and essential for today. It is fair to ask, what would the world be like today if the core goodness of each of these faiths had never existed? I believe that, with all of humanity's shortcomings, misdeeds, injustices and selfishness, it is the basic teachings about love, sharing, caring, fairness and forgiveness, taught and practiced by Jesus Christ, that constitute the best and greatest thing to ever happen to humanity.

The Last Supper of Christ *by Guillaume Herreyns (1743 - 1827).*

VIII.
Dualities of Existence

Evolution of Everything

As we seek an interconnection between science and religion, the phenomenon we know as "evolution" enters the picture. Evolution as a process has its own evolution in the origins of how we understand it. There was that ancient time when Aristotle declared that whatever existed at that moment in time was the result of a directive emanating from an out-of-this-world God, a notion which required more than two thousand years to be challenged and changed.

It was not until the 1800's that natural scientists like Jean de Monet, Robert Chambers, and Charles Robert Darwin began probing more deeply into what came to be known as "The Origin of the Species." It was in 1859 that Darwin wrote his famous and provocative book proposing a theory of evolution by natural selection, out of which emerged the theory of the survival of the fittest. Not until more recent times did scientists conclude that all living things first began with their appearance around 3.5 billion years ago as a one-celled organism, a basic form out of which all living matter of today is descended. From the Big Bang, until this very moment, all existence is still evolving, be it the cosmos, or microbes or humans. There is the notion among some that even their God must be evolving. That is, in whatever form God may or may not exist.

Charles Darwin,
Age 45, 1854

Dominance of Dualism

The discovery of the birth of living things, and the evolutionary development thereafter is only part of the story of life and of existence. There is yet another phenomenon that just may be the most puzzling embodiment of all existence.

I am referring to the significant phenomenon which I define as "the dualism of existence." Observe how so much of existence comes in pairs. Such a phenomenon becomes all the more puzzling when we remember that all living matter began in the form of single cells. Let's examine that which is closest to us, the human body. Count the duality, the dominance of pairs in our bilateral symmetry: our eyes, ears, nostrils, arms, legs, lungs, kidneys, testes, ovaries, even the pairing of our right brain and left brain. Look outward to our societal existence: birth and death, male and female, yin and yang, yes and no, positive and negative, left and right, up and down, in and out, high and low, healthy and sick, rich and poor, young and old, good and evil, conservative and liberal, happiness and sadness, and so on.

Then, there is the binary life around us: plants and animals, land and sea, day and night, hot and cold, north and south, east and west, creation and destruction, still and moving, positive ions and negative ions, sending dendrites and receiving axions, and more. Even humanity's most amazing creation, the computer, and all of its byproducts, operate on a dualistic code of "zeroes and ones." For further confirmation of this pervasive dualism, consider macro and micro science: stars and planets, light and dark, hot and cold, matter and energy, magnetic attraction and repulsion, protons and electrons, astrophysics and nanophysics.

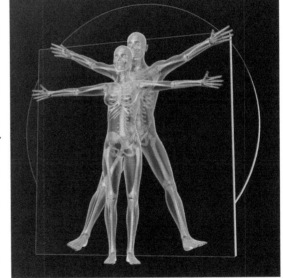

Dualism as exhibited by an adaptation of Leonardo Da Vinci's, The Vitruvian Man.

There are still two more dualities which deserve our attention. While we are keenly aware of the difference between the male and female of our human species, there follows the fact, the mystery, that there is almost an equal number of human males as there are females on this earth. More specifically, United Nations population experts have concluded that our earth as of 2012 is populated by 3,487,869,561 human males and 3,439,427,037 human females, a ratio of 50.4% vs 49.6%. This appears to be just one example which suggests that dualism may have a special purpose: "balance."

While some may argue that this all-prevailing dualism is needed for balance, or serves in some way as a necessary function of all existence, the serious thinker should look more deeply. Especially important is a further consideration of one particular dualism—the problem of good and evil and of pleasure and suffering.

Morality of Duality

Just as Charles Darwin and others felt obliged to understand how the origin and evolution of the species worked, we are equally compelled to understand one particular matter, that of the duality of almost everything. Buried in the midst of all this duality is one factor of paramount relevance to the lives of human beings, and of paramount relevance to the religious beliefs of those humans. It is the duality of good and evil, and of pleasure and suffering.

Human beings have evolved through hundreds of millions of years, from clusters of single cells to the Homo-sapiens of today. With this evolution has come the evolving of the human from a being conscious of its existence, to a being subconsciously mindful of other things beyond itself, to a being conscientious about what needs to be. That seems like progress, and perhaps a promise of better things to come.

All that is well and good, but what about those most puzzling of all dualities, good and evil and pleasure and suffering? Does the dominating presence of the duality of almost everything mean that evil and suffering are simply a natural and given factor of life? Does this mean that, realistically, humans are stuck forever with the scourges of evil and suffering? Does this mean that all the disparities listed earlier are simply an unconditional given of cosmological existence and of human life? At this time in the evolution of the human mind, we simply do not yet understand the purpose or the morality of dualism any more than we understand the purpose or origin of existence. I believe that there is reason for us to be patient, rather than for us to be dismissive or despondent. Here is an example of when we do not yet know the answer to something important, then there may be good reason for us to simply hope.

Evil with a Purpose

Several years ago, while on a family winter skiing vacation in alpine Austria, I experienced something I will never forget. It was a shivering cold, snow blanketed night at our hotel, and I simply couldn't sleep. Restless, I dressed and went down to the hotel lobby, where I discovered the night clerk on duty at the front desk.

Eager for some kind of dialogue worthy of the lateness of the night, my inquiry revealed that my new night-bound friend was a one armed, hard looking, former Nazi soldier. Of course, this peace loving American that I considered myself to be eagerly asked some probing questions of my captive Nazi soldier. The memorable message of the evening is one I will carry the rest of my life. "Yes, I lost my arm in a battle, but it was my contribution to the war. The Germans saw the truth that the Allies never understood. Wars are necessary for the advancement of humanity. Wars are the means by which you cleanse the world of the weak and the worthless, a fulfillment of the purpose of evolution, that is: the survival of the fittest. Wars are an absolute necessity for the advancement of humanity!" Never having heard before this explanation of the Nazi cause, I recognized that any rebuttal I would offer would make no difference in the war hardened mind of this committed Nazi veteran.

I returned to bed, disturbed by the message of my hotel mate, and no little bit anxious about my old friend Charles Robert Darwin. Then, as I returned to bed and pulled up the covers, I remembered that it was Christmas Eve. What a contrast and what a duality, a story of war and suffering set in the midst of a time of peace and love. Here was living proof that there are both physical dualities and moral dualities, and evolution has uniquely issued both to the human being.

Evolution of Dualities

Remember that astrophysicists have determined, without doubt, that our Cosmos, or Universe, is expanding, and doing so at an exponential rate, and that it has been doing so ever since its birth 13.8 billion years ago.

Likewise, the ongoing work of our learned particle physicists has led to an almost explosive discovery of more and more of the tiniest things which make up our basic component of existence, that is, the atom and its elusive family. In previous chapters, we have even noted that the world's population and knowledge and wealth are expanding at an exponential rate. But, I suggest that not everything is expanding. There is another interesting duality. It pertains to the shrinking of our world.

The Shrinking World

The visiting speaker was known to all as a remarkable guru, possibly the world's most brilliant economist. Officially, he was the Chairman of the Federal Reserve Bank of the United States. The occasion was a tribute to the great work and deserved retirement of an esteemed mutual friend.

This was the auspicious occasion of the retirement of The President of the Atlanta District of the Federal Reserve Bank of the United States. So, inspiring remarks of gratitude from a very special keynote speaker were in order. But, this was not to be. The honored guest and keynote speaker was indeed a guru. You knew it the moment he began, so soft and carefully spoken that the crowd could hardly hear, and so fast talking in technical terms that his listeners could hardly follow.

The speaker was the much esteemed and honored Alan Greenspan, the person to whom the world's financial order was dependent on his every word and nuance. His topic was not an insider's revelation of interest rate changes or market shifts or cycles to come as his audience had eagerly anticipated. Dr. Greenspan's topic of the evening was "The Miniaturization of Society." What a disappointment, I thought. No insider secrets to help guide tomorrow's investing world? Little did we expect that we were about to learn something far more important, one of the most important changes taking place in the history of the world. Dr. Greenspan's message left his mesmerized audience clutching to every word of his incredible story of the "rapid shrinking of the

International Space Station with astronauts over planet Earth.

world." Greenspan was not talking about money, he was talking about something entirely different. In a change-filled single century, he reported, society was experiencing an exponential shortening of the world's space and time.

As to space, travel since 1900 has shifted from the slow moving ocean liner's travel at 20 knots per hour, to the early automobile's travel on land of 30 miles per hour, to the first airplane's travel at nearly 100 miles per hour to the jet airliner's travel of today at 600 miles per hour to a manned rocket's travel to the moon at 25,000 miles per hour. In other words, wherever you may wish to go today, you can cross that distance in a mere fraction of that required at the beginning of the 20th century. Simply put, the world has dramatically shrunk, opening the way for us to travel more quickly, more distantly and more often. Surely, these changes offer enormous benefits, accompanied by enormous adjustments.

As to time, communication since 1900 has shifted from telegraph messaging on wires strung along railroad tracks, to telephone talk via personally staffed switching stations, to under-seas cabling, to micro-wave relay towers, to internationally broadcasting television, to globally inter-connected satellite systems. In other words, with whomever you wish to communicate, be it an individual or with masses, we have gone from complicated methods and slow moving messaging to information received within seconds by unseen millions located all around the globe. As to time, the world has dramatically shrunk, opening the way for more people to know more instantly about more subjects.

As to process, communication of 1900 has shifted from the simplicity of a hand written letter, to a hand pecked typewriter, to a mimeograph machine, to a Xerox copier, to a room-filling Remington Rand main frame computer, to a fax machine in every office, to the magic of a desktop computer, and on to the portable laptop as a standard around the world.

The occasion of Dr. Greenspan's talk took place in the late 1990's. Think how much more the world has shrunk since then: orbiting space stations, television sets owned by most of the world, and communication around the world packed into a smart phone the size of a hand. Then, a company with the funny name, "Google," has come along and placed at our finger tips an instantaneous encyclopedia of information about anything and everything. Next, employing the earth-circling 24 satellites of the United States Defense Department, that same company has brought to our fingers a means of instantly locating almost any place on Earth, and the path of how to get there from here.

Now that same crazy named company and others are introducing the technology for a driverless vehicle to drive itself without the need for human control. Think, too, about all the other technological ways of staying connected: Facebook, Twitter, YouTube, SnapChat, Skype, Facetime, Instagram, and Alibaba, with more announced almost daily.

Reasons To Rethink

It is comforting to know that the march of evolution is bringing to us comforts and capabilities never dreamed before. It is disturbing, however, to know that the march of evolution is still leaving us with suffering, injustice, war, atrocities, murder, hurt and wrong.

If existence is so miraculous and if God is so great, then why all the terrible? If existence is so miraculous and if God is so great, then where is the proof that there is nirvana after our death? If existence is so miraculous and if God is so great, then why is there the hypocrisy of dualism?

Are we to accept the cold-hearted philosophy of that one-armed, war-hardened Nazi soldier? Are our deep-felt fears to be eased by the simple proclamations that "Jesus Saves," or that "Allah Blesses," or that "Buddha Knows?" Even with the invention of all the miraculous devices for the detection and conveyance of information and the discovery of all the miraculous workings of the Cosmos, factual and provable answers to the problem of evil and the question of eternity continue to elude us.

This much does seem clear. If we allow our lives to be totally consumed by anxieties about terrible harm and evil, and about absolute assurance of an afterlife, we will only crowd out the possibilities for an existing life of happiness and purpose. In the meantime, find some comfort that the recent discoveries of Big Science are offering some positive possibilities as to the basic nature of all existence, including that of you and me.

My personal experience is that a preoccupation or obsession with the negatives and uncertainties of life are destructive to the power of the goodness which lies within us. Instead, a preoccupation with the serving of others is the key to unlocking happiness for ourselves and for those whom we may serve. Simply put, being filled with the "giving away of ourselves" is the means by which we may best deal with evil, suffering and death.

I like best the challenge of my dear friend, Bishop John Shelby Spong, ordained Episcopal priest, eminent theologian, brilliant scholar, compelling lecturer and prolific author. Simply put, Bishop Spong offers as life's best advice for the best way to live: "Above all else, love wastefully!"

IX.
Miracles of Existence

Faith Based On Facts

As if all existence began only a few thousand years ago, the Judeo-Christian Bible offers multiple reports of miracles which claim to be original, unique, and exclusively true. Remember the stories of the parting of the Red Sea, Moses and God's Commandments, Noah and his ark, Jonah and the whale, God's promises of eternal life, and God's warnings to believe. Later writers, following the time of Jesus, declared Jesus as God's only son, further insisting on the healing of the sick, walking on water, multiplying a quantity of food, promise of eternal life, raising from the dead, ascension into heaven, and so much more. Surprisingly for some, similar stories are found in other faiths. Remembering the reality that such stories are recollections of writers who lived and wrote 50 to 100 years after the sincerely believed events would have taken place, we need to be very careful before we employ these stories as the factual basis of our belief and as the impervious core of our religious faith.

It is the present day courageous research and teachings of progressive thinking American theologians like John Shelby Spong, Dominic Crossan, Karen Armstrong and the late Marcus Borg, which have brought to light in modern times many of the misinterpretations and self-serving messages coming to us from ancient times. Perhaps these ancient stories and proclamations were well-intended claims based on dreams and hopes, but to the genuine scholars of today these were hardly factual accounts. Conversely, there are some real miracles taking place all around us every moment. They are best defined as *the miracles of existence*. And, there is a universally true, factually proven, life-worthy reality to each. They are the life-guiding messages to be found in the factual discoveries of astrophysics, nanophysics, particle physics and neuroscience, what I have come to define as "Big Science."

My Awakening

My own understanding of science was first formed by the blurred and blinking lights of those stars I could see in my *Popular Science* magazine-advertised telescope, and my viewing of blood cells with my F.A.O. Schwartz store-bought microscope. My curiosity was piqued far more when, as a teenager, I was privileged to attend the celestial showing of the Hayden Planetarium in New York City. Out of these simple experiences, I gained a passing knowledge of, and an insatiable curiosity for, what science was all about, but when I was confronted with the mandatory classes of physics and chemistry in college, I quickly recognized that I was not made to be a scientist, and certainly not a physicist or chemist or biologist. For the next several decades,

The Hubble Space Telescope in orbit above the Earth.

my teenage interest in things scientific shifted into dormant mode.

It was that innocuous and simple announcement in our church bulletin about my friend's talk about outer space, which I credit for waking me from my scientific hibernation. How glad I am that I chose to be on hand for what proved to me to be a fascinating and important revelation of what really goes on far out there in distant space.

I still carry with me that gloriously-colored, dreamily-floating, strangely-beautiful image flashed on the screen, a hauntingly real depiction of a newborn star appearing in the nebula HH 47 at a distance of 1,500 light years away and discovered by the Hubble telescope orbiting outside the atmosphere of the Earth. This was a depiction of the birth of just one star, and I was reminded that there were still more than a hundred billion times a hundred billion of them. I was awestruck, humbled and puzzled. Since that memorable morning, I have learned there is so very much more to learn, to contemplate, to understand.

What followed was an insatiable appetite for me to read every book I could and to witness every lecture available, all with the determination that I try to understand more fully the miraculous workings of our Universe, and indeed our very existence.

Big Science Discovered

Famed Astronomer and Astrophysicist Carl Sagan

Along the way to my "born again" experience and understanding of matters scientific, know that I am especially mesmerized by, and grateful for, the Big Science research and presentations in the 1980's of the late Carl Sagan and his wife Ann Druyan, extraordinarily talented people who have offered to the modern world a clearer understanding of the workings of the Universe, its composition and functioning, its mysteries and meaning, its history and future—a summary education for all the world. Their all-time classic book is *Cosmos*, and their all-time classic television series, *Cosmos, A Personal Voyage* was widely broadcast in 1980.

I am equally grateful for the extraordinary production and presentation of television's 2014 thirteen part program, *Cosmos, A Space Time Odyssey*, produced under the leadership of Ann Druyan and hosted and narrated by Dr. Neil deGrasse Tyson, Director of the Hayden Planetarium in New York City and Professor of Physics at Columbia University.

The *Cosmos* programs bring together astronomy, astrophysics, quantum physics, biology, botany, geology, anthropology, neuroscience, and history, as never before assembled and offered by anyone. These two programs present for all an essential, factual and clear explanation of what our existence is about, knowledge which should be a required offering to every high school student in the world.

While these astronomers and astrophysicists have brought to the world an understanding and appreciation of the wonders and workings of the Universe, and, in fact, all existence, it may come as a surprise that His Holiness the Dalai Lama, the world leader of the Buddhist faith, authored in 2005 a compelling original work about the

The 14th Dalai Lama, leader of the Buddhist faith.

Universe, entitled *The Universe In a Single Atom, The Convergence of Science and Spirituality*. Here is a world famous religious leader who also understands and believes in the need for the public, in general, and the religious, in particular, to know, embrace and incorporate Big Science into their personal faith. Interestingly, this writing and its important message came from a person of a lesser known faith with a lesser number of followers, and not from any leader of the far more dominant and influential faiths of Christianity and Islam.

The Awesomeness of Astrophysics

It is likely that it was Australopithecus of 4.5 million years ago who first looked up and viewed with both wonder and fear the twinkling lights in the night darkness of the sky above. Little did this human ancestor understand that those stars had been shining for billions of years before the earliest form of humanity evolved into being. It took 1.5 million years for early human like beings to advance from using bone tools to stone tools. Next, the earliest forms of Homo Habilis appeared around 200,000 years ago in the Rift Valley of today's Kenya and Ethiopia. Over a time of around 100,000 years those earliest Humans gradually migrated up into the Mesopotamia area of the Middle-East. From there, Homo Sapiens migrated out across Europe and Asia, followed over the next 10,000 years with migration over into the Western Hemisphere, that is, the Americas of today. Then, to give history its due, it was only around 600 years ago that Michelangelo was commissioned to paint the Creation of Adam by that grandfatherly God figure reaching down from Heaven.

Billions of years after stars were born and scattered into space, in the year 1920 CE, atop a mountain in California, a man named Edwin Hubble focused a hundred-inch telescope on the night sky, leading to a succession of miracle-like discoveries which continue to this day. Hubble and his team are credited with leading later scientists to the discovery that the Universe is subdivided into hundreds of billions of galaxies, each of which, in turn, is comprised of as many as hundreds of billions of stars.

Possibly even more surprising is the discovery that all these galaxies and all their stars are rushing away from each other at an exponentially accelerating speed, and that they have been doing so ever since the "existence birthing" of the Big Bang explosion nearly 13.8 billion years ago.

To better understand and appreciate the scale of our cosmos, note that the distance between our earth and our sun is 93 million miles, or approximately 150 million kilometers. The distance from our Sun to its next nearest star is 24 trillion miles, or 40 trillion kilometers. What then does all this say about the size of the Universe?

Because the Universe is still expanding exponentially at a rate (in space/time) faster than the speed of light, and remembering, also, the speed of light is 186,232 miles per second, or 670,615,200 miles per hour, it is virtually impossible to catch the Universe for a moment in time long enough to measure its size. The best assumption is that the observable Universe, that which we can detect at this moment, is 558 billion trillion miles or 883 billion trillion kilometers in diameter. Because we cannot measure space beyond the edge of the observable Universe, we do not know whether the Universe is infinite or finite.

Now with the advancement of the more sophisticated space-based, solar-orbiting Kepler telescope, and earth-orbiting Hubble telescope, more essential and exciting discoveries have been rolling in. Then, with the forthcoming James Webb telescope,which is due for activation in 2018, there is no telling what other amazing discoveries await us.

The Universe began in an explosion of pure energy, an event popularly called the "Big Bang," an event happening about 13.8 billion years ago, an event that took place in less than a trillionth of a second. The original explosion has been expanding and creating ever since. The light coming from the farthest stars we can presently detect has been on its way to our planet for about 13 billion years, traveling on its way some 77 billion trillion miles. We also have come to know that the instantaneous speck of light from a star detectable by the naked eye has been on its way to our planet for many thousands of years. Further, we have determined that the speed of light is 186,232 miles per second, or 670,615,200 miles per hour, and as far as we can tell, nothing else in all existence moves faster than the speed of light.

The work of astrophysicists is continually explaining more and more about the life and death of the Universe and everything in it. We now know that its existence is evolving, that the Universe functions, like humans, in cycles of birth, life, and death. That means that the Universe in which we exist will not exist forever. We have good reason to wonder if there was an earlier Universe which gave birth to our present Universe. As if that isn't a curiosity enough, there also are astrophysicists conducting research to determine if there may be such a thing as a multitude of Universes.

In a parallel way, our planet is going through its own evolution. Earth came into existence about 4.5 billion years ago, long after the birth of the Universe. Earth has been going through its own struggles for survival and advancement, to draw upon an analogy to the evolution of humanity. Earth has been cooling from its super-heated beginning, but also is predicted to ultimately burn to a crisp, lifeless rock by the overheating of our sun, followed by the burning out of our sun. Interestingly, the very core of our earth is still burning and churning at a temperature around 10,000 degrees Fahrenheit or 6,000 degrees Celsius.

In the meantime, our Earth has been, and continues to be, bombarded by asteroids, radiation, and who knows what else. It has experienced the coming and going of continents and oceans, of the eruptions of volcanoes, of the sliding of tectonic plates, of ice ages and dinosaurs, of life-giving photosynthesis and life-sucking evaporation. Incredibly, we also now know that all living things, including us, originated from single cells.

As with the Universe, there will come the time of universal death, except that it will happen on Earth before it happens to all the Universe. Scientists speculate that life on Earth will be around for another few billion years, if not first destroyed by a cosmic or geological disaster, or a human wrought disaster. Just think, in spite of these fears, we would be well advised to believe there is still time enough for humanity to get its act together.

While NASA scientists are trying to determine if there could be life on other planets, cosmologists are busy theorizing about whether there could be other universes and what they could be like.

Before we conclude that we humans now know it all, physicists quickly remind us there is so much more to be considered about our existence, and so much more that is beyond our current human knowledge. As proof of how much there may be beyond our full comprehension, we turn to a listing of astrophysics terms published in a 2015 edition, *Scientific American.* Here we learn of the ongoing research of a strange-sounding list of items, such as black holes, dark energy, dark matter, antimatter, cosmic inflation, space-time, spiral galaxies, elliptical galaxies, neutron stars, vibrating-string theory, supernovae, electromagnetism, quantum gravity, firewalls, wormholes, pulsars, cosmic rays, singularity, super-symmetry, standard candles, event horizon, quantum entanglement, quantum fields, higher dimensions, Planck scale, and much more. This does not include all those unknown, unnamed phenomena, yet to be discovered. With all this, how can we not be in awe of our existence?

What do we make of this beyond-belief, incredible, miraculous existence, and our human place in it? If we do not yet know all the answers, there is still some convincing evidence that, in the meantime, we humans would do well to go ahead and live our lives in awe and allow humility to be our guide. And while we continually seek perspective and understanding, there is still so much more to be thinking about.

The Boundlessness of Nanophysics

From the vastness of the Universe, consider the opposite. Consider now the smallest of that which makes up the Universe. To put "small" in perspective, start with the fact that the head of a straight pin is comprised of millions of iron atoms. Put another way, just one grain of sand on a multi-mile beach contains more atoms than the total number of grains of sand on that entire long beach.

Just as the Hubble telescope floating far above Earth is revealing the specifics of that which is still much farther beyond, the Large Hadron Collider

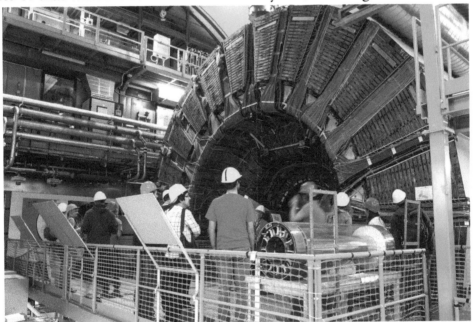

Large Hadron Collider at CERN Laboratory, Geneva.

buried below ground at a site outside Geneva, Switzerland is revealing the details of that which exists deep inside that which we see, feel, and understand.

Operated by a consortium known as the European Center for Nuclear Research, or "CERN," a massive assemblage of equipment placed inside a seventeen mile circular tunnel is constantly smashing atoms and examining the particle debris—another of the amazing devices produced out of the inventiveness and curiosity of the human mind.

While atoms may be defined as the basic units of all matter, matter is comprised of even smaller particles and charges. Those subatomic particles are defined as elementary particles, and these smallest of the small are various combinations of protons, neutrons, electrons, quarks and neutrinos, and most recently verified, the still smaller Higgs boson.

Interestingly, neutrinos and bosons can operate independently and separately from the atom. In fact, the neutrino is so elusive that it has been nicknamed the Ghost Particle, and the even smaller and more elusive Higgs boson has been nicknamed the God Particle. The size of these is so small and elusive that there is some claim that they are a form of energy, rather than a form of matter.

Discovered in the 1950's, but accurately measured in 2014, the story of the neutrino is a truly mind-blowing revelation. Every second of every day, trillions times trillions of neutrinos are zipping through Earth and everything on it, racing at the speed of light while rarely even touching any atom as they pass through us, or even as they pass through Earth. This also means that there are trillions of neutrinos zipping through us every second.

Just a few years ago The CERN laboratory in Switzerland fired a stream of neutrinos from Geneva, Switzerland to Gran Sasso, Italy, a distance of 450 miles. What is significant about this is that the neutrinos passed, not over, not around, but through the solid rock Swiss Alps, traveled at the speed of light, and reached the receiving station in Italy in a trillionth of a second. In other words, these strange and ghostly "things" traveled unimpeded by so-called impervious solid rock mountains and their composition of molecules and atoms, racing through, unhindered, at the speed of light. As if to reconfirm these findings, a similar experiment was performed in 2014 by Fermilab near Chicago, and did indeed reach the same conclusions.

What we humans see is not what we think we see. Empty space is not empty. It is busy, crowded, and what is in it is elusive. What is out there and what is within us is an unimaginably vast number of unimaginably small, unseen, unfelt and unheard swarms of what can be best described as "ghosts."

Could it be that those ghostly particles are doing more than just racing around on a joy ride? Could it be that they have an unseen, unmeasured function yet to be determined? Could they have some connection with electromagnetic waves or gravity? Could they be playing an important role in existence, a role we are yet to understand?

These most recent mysterious and mystical discoveries of science give reason for humans to admit that there is much, much more out there than we can even begin to imagine, that there is much, much more yet to be discovered, and there is so very much we do not yet know, or understand. Perhaps all this suggests that we share our existence with an immeasurable and undetectable number of "strange friends," with whom we should try to be friendly, respectful, intrigued, and strangely, thankful.

The Connectedness of Particle Physics

Physicists now report that there is another supremely significant matter for all society to take into serious consideration. Whether it is the farthest star residing an immeasurable distance from planet Earth, or a planet circling in our solar system, or the highest mountain on planet Earth, or water filling the ocean, or leaves on a tree, or a fish swimming in a lake, the hairs on our bodies, spermatozoa on their way to an egg, or the neurons flashing in our brains, everything and everyone is made of the very same set of three things: up-quarks, down-quarks and electrons.

Now, here is the real message: If everything in all existence is made of the same parts, then, whether animate or inanimate, everything in our cosmological existence is a "kissing cousin," one to the other, ultimate proof of true universality—an interconnecting kinship among *all* existence. Here is profound proof and reason for *all* humans to care and share and respect each other, and to appreciate and conserve and protect all that surrounds and sustains us.

Perhaps this proves that there is preciousness about all existence—a preciousness not to be wasted or destroyed or ignored. Clearly, war, violence, and waste are the ultimate violation of all the laws of nature, a crime against all existence. Clearly, human caring and sharing and compassion are the ultimate fulfillment of all the laws of nature.

The Boundaries of Neuroscience

As if the workings of outer space and inner space were not complicated and miraculous enough, the workings of the human brain may be even more amazing. Neuroscientists remind us that, of the trillions of cells which make up the human body, the brain has operating in it between 100 and 200 billion neurons. More amazingly, the number of ways that information flows in and around and among the neurons of the human brain is more numer-

Neurons in the Brain

ous than all the stars to be found in a galaxy. It is as if our human brain is a cosmos all its own.

Because they are all comprised of subatomic quarks and electrons, there must be an interrelationship among the mysterious neuroscience of the busy human brain, the immeasurably vast astrophysics of the Universe and the inconceivably small particle physics of that which makes up each. That there is interconnectedness among these three levels of reality appears to be the fundamental basis of all existence.

The factual reality is that, whatever there may be, the human understanding of it is limited by the boundaries of what the human brain can perceive. To put it another way, we perceive no more than what our brains can detect and process, and we can know only what is in the range of the human brain to perceive. The real and truthful conclusion is that we humans do not know everything, although we all too often act as if we do.

The positive news is that just as the Universe is ever expanding, so are the limits of what the human brain can perceive, and, with it, what the human mind can conceive. Look at this expansion in another way. Some of the evidence of the earliest existence of Australopithecus, our earliest tool-using ancestor, and living much earlier than better known Homo Habilis, dates back to over three million years ago. In the evolution of our ancestors, paleontologists have found that it took 1.5 million years for Homo Habilis to advance from using natural bone tools to using sharper stone tools. Then, in only the past 50,000 years, Homo Sapiens has advanced from its earliest invention of cave drawings and figurative carvings to such scientific and technical marvels of today as jet airplanes, nuclear energy, lung replacements, space travel, and smartphones. This comparison offers convincing evidence that, as the Universe is expanding ever faster, so also is human cultural and biological evolution speeding faster.

Reasons for Awe

If existence is so incredibly vast, and at the same time so incredibly small, and at the same time so incredibly complex, where in all this do humans fit? There was that time before Stephen Hawking, before Edwin Hubble, before Albert Einstein, before Isaac Newton, before Galileo Galilei, before Nicolaus Copernicus, when life appeared to be simple. The scope of human perception back then was limited to belief in a flat earth, to the distance a person could walk or sail, to a reigning God residing in an imagined heavenly kingdom, to a warmth giving sun nearby and to the light giving stars above.

It was in 1508 that Pope Julius II of the Catholic Church asked artist Michelangelo to demonstrate more about this beautiful simplicity by depicting on the ceiling of the Vatican's Sistine Chapel an illustration of a human-like God giving birth to Adam. With that painting, Michelangelo gave humanity what was then thought to be a complete, virtually total, explanation of what life was all about. Keep in mind that this depiction conveyed what was universally believed only 600 years ago, and, because this depiction was so simple, beautiful and compelling, it still defines the religious faith of billions of people today.

Now come all these truly explosive scientific revelations of what existence is really all about, a revelation seemingly ripping apart what humanity had "known" for thousands of years to be the "God-Given-Truth" about our world and about humankind's place in it.

Our Milky Way Galaxy as seen from Earth.

In response to the limitations, changes and differences posed above, I propose to all a universal acceptance that we rethink our views of life and of existence, turning now to a fact-based faith of awe, wonder, hope, thankfulness, humility, and compassion, rather than holding to a supernatural-based faith based on a human-looking god, reigning over earth from a place in Heaven, promising forgiveness for our sins and a neverending life in Paradise. While a fact-less supernatural kind of faith may constitute the most popular beliefs of these times, they are too often the very cause of the divisions and disputes which prevent our world from being what it can be.

X.
MYSTERIES OF EXISTENCE

Why Not Nothing?

There is one question about our existence which seems to trump all others. It is simply: "Why is there not nothing?"

In his recent book, *A Universe from Nothing*, discussed on National Public Radio on January 13, 2012, physicist and philosopher, Dr. Lawrence M. Krauss puts it one way: "Why is there something, rather than nothing?" This, then, leading to another question: "Has our existence been born out of nothing?"

In a May 14, 2013 article in *The Washington Post*, Science Columnist Joel Achenbach states the matter in a personal context: "In the late 1990s I made a list of the five biggest unanswered questions in science. All obvious stuff, like how did life originate and how does consciousness emerge from the brain. But number one, the foremost mind-boggler, the ultimate question, was, why is there something rather than nothing?"

In his 2004 book published by Clarendon Press, Oxford philosopher Bede Rundle also questions, "*Why is there Something Rather Than Nothing,*" and then concludes that "There has to be something."

My own take is that if there were nothing instead of something, we would not even exist to be able to pose a question of asking "why," or "if," or "whether" about anything. As long as there is something there cannot be nothing.

Why do we even bother to pose this question? I believe that if we are going to address matters as basic as all existence, then we need to start with a clean slate, a blank page, and then initiate our probing of existence from there.

Who, What and Where Is God?

In some form or another, and by a variety of counts, most of the population of the world believes in a "God"—an all-powerful and all-knowing being existing somewhere within or beyond known existence. For most of the "believers," God includes and incorporates human characteristics, while also existing beyond humanity. We have come to give this above-and-beyond-it-all-being, this incredible and mysterious force, as elegant and as exquisite a name as our imagination can carry us: "Divine."

As already established, there are a variety of interpretations and beliefs as to the summary description of God and even as to the very existence of a God. They range from the Theist, to the Pantheist, to the Panentheist, to the Agnostic, to the Atheist, to the Humanist, and variations thereof.

While whole nations, like China, may officially reject the notion of a God, and while the 2014 *U.S. Religious Landscape Study* of the Pew Research Center demonstrates that there is a gradual trend among most of the world's newer generations that they have little or no interest or belief in a God, there still remains today a larger share of those who "believe."

The British publication, *The Independent*, ran an article on March 10, 2009 in which several scientists were quoted as to their explanations for a belief in God. Professor Jordan Grafman, a neuroscientist, writes, "Religious belief and behavior are a hallmark of human life, found in all cultures, and with no other animal equivalent." Dr. Grafman goes on to state, "When we have incomplete knowledge of the world around us, it offers us opportunities to believe in a God. When we don't have a scientific explanation for some-thing, we tend to rely on supernatural explanations."

On the other hand, it is essential that we also look at the antithetical. In these times, there is one "expert" who best embodies the characteristics of an "atheist" or "non-believer." That expert in non-belief is popularly recognized as Richard Dawkins, Professor of Public Understanding of Science at Oxford University, whose 2006 book, *The God Delusion,* is widely acknowledged as offering the most profound argument against the existence of God.

In his book, Dawkins argues, "Imagine, with John Lennon, a world with no religion. Imagine no suicide bombers, no 9/11, no 7/7, no Crusades, no witch-hunts, no Gunpowder Plots, no Indian partitions, no Israeli/Palestinian wars, no Serb/Croat/Muslim massacres, no persecution of Jews as 'Christ-killers,' no Northern Ireland 'troubles,' no 'honour killings,' no shiny-suited bouffant-haired televangelists.... Imagine no Taliban to blow up ancient statues, no public beheadings of blasphemers, no flogging of female skin for showing an inch of it." And, that is just one small part of Richard Dawkins's encyclopedic set of arguments that "God is a Delusion."

We must acknowledge another scientist who also rejects the notion of a God. The late Dr. Stephen Hawking, Professor of Mathematics at the University of Cambridge, is generally regarded by the scientific world as one of the most brilliant theoretical physicists since Albert Einstein. Hawking is the author of such advanced thinking and writing as found in his *A Brief History of Time,* and in *The Universe In A Nutshell.* Dr. Hawking's mission appears to be to uncover the Holy Grail of science, the elusive "Theory of Everything." Along the way, he insists that God is the mere faith invention of the human mind, an invention not verifiable by scientifically proven fact. So with no god, where does the genuinely concerned, deep thinking, morally good person turn for the truth?

My response is: Why can't we simply, but fervently, believe there is a dimension to existence which transcends all that we humans now know or even imagine. And if that dimension surpasses our human perception, then that is still acceptable, because it allows us to hope and believe in possibilities not yet present or measurable to us, but nevertheless possible. I suggest that as long as the human condition falls short of what needs to be or what could be, there is good and fair reason that we humans should imbue ourselves with a hope for something better and with a belief that answers may lie beyond our knowing.

Believe that there may be a dimension of existence which goes beyond our factual knowledge. If you prefer, you can call it God. Just don't belittle and demean this God by attributing to it such solely human characteristics as: "father," "holy," "love," "forgiveness," "acceptance," "rejection," "salvation," "condemnation," or even the term "creator." Be in awe of, and humbled by, that which may exist beyond us, but do not kidnap that which is beyond us to be our personal savior against that which may harm us, be it now or after we die. Believe and hope, but always first let facts and reason carry us as far as human knowledge allows and as far as our human brain perceives. In the believing, acknowledge that we just may not know everything.

The Simultaneity of Nature

Just outside my office there are three supremely graceful and majestic pampas grass bushes. They are at least 25 yards apart, but that isn't the point. In the spring, on exactly the same day they come alive with new shoots of fresh greenery. But, today, just today, only today, their beautiful, feathery white seed heads have burst into being with all the beauty which nature can muster, like an array of New Year's fireworks, symbolizing a celebration of a memorable place and time. Why now? Why exactly the same time? Why exactly the same look?

The same could be said of a flock of sky-bound geese, gathered in a precise V- shaped formation, dutifully headed south in the fall and north in the spring. We note too that, in the springtime, the leaves of the trees and the blossoms of the flowers come alive in their fresh glory, in unison and in harmony, as if together singing Handel's *Hallelujah Chorus.* Why so perfectly in unison? Why so precisely timed? Why so exactly tied to the same time year after year?

Do the flowers know how to speak to the kindred of their species? Do the birds know how to talk to their cousins? Do the trees know how to speak to their neighbors? Do they all carry their own version of a calendar and a watch? Does the rest of nature know something we do not?

For now we must accept the fact that we really do not know. Instead, we would do well to enjoy and be thankful for the mysteries and miracles and wonder we so constantly experience.

The Centrality of Humanity

Dr. Joel R. Primack is a distinguished Professor of Physics and Astrophysics at the University of California Santa Cruz, and is the co-author of the 2006 published book, *The View from the Center of the Universe.* This learned scientist offers evidence of a fascinating claim: "The human being just happens to reside right in the middle between what is the largest of what we know and what is the smallest of what we know."

Explaining more fully, Primack writes: "From the Planck scale to the cosmic horizon, the Universe encompasses about 60 orders of magnitude—60 orders of magnitude separate the very smallest from the very largest." Using the "Cosmic Ouroboros," (a snake like being circling around to swallow its own tail), an illustration created by Dr. Sheldon Glashow, Primack goes on to explain, "Traveling around the serpent from head to tail, we move from the scale of the cosmic horizon to that of a galaxy super-cluster, a single galaxy, a solar system, the Sun, the Earth, a mountain, a human, a single celled creature, a strand of DNA, an atom, a nucleus, the scale of weak interactions, and approaching the tail, the extremely small scales of dark matter particles such as the axion."

Put a simpler way, if we start with the height of a human, and multiply this dimension by 10 to the power of 30 we are at the scale of our entire Universe. Conversely, if we start with the height of a human, and multiply this dimension by 10 to the power of minus 30, we are at the scale of the smallest thing we know to exist.

So, what does all this sophisticated scientific jargon really mean? It means that we human beings exist at the very dead center of all known existence. If medieval astronomer Copernicus were alive today, he would likely exclaim "there we go again." It was Copernicus who first proclaimed that the Sun, and therefore the then known cosmos, does not circulate around the Earth, but the Earth around the Sun.

That is just a reminder that these astrophysical revelations, illustrated by the "Ouroboros," do not suggest that the Universe revolves around the human being, but it is still exciting, and a bit scary to think that we, the human beings, are sort of "existence's centerpiece," What does that mean?

Was the universe created just for us? That is a flattering thought, but there is no provable evidence to this effect. Does our central position among everything else give us a special responsibility? That, too, is an interesting thought, but again one for which we have no proof. Does this human centrality carry any significance? Who knows, but the notion is certainly worth pondering.

In the meantime, until we know differently, we humans would be well advised to go ahead and live out our lives humbly and gratefully, and do so as if we are indeed the centerpiece of existence, benefiting from what appears to be a supremely beneficent existence, but also responding to what appears to be an equally responsible existence.

The Uniqueness of Earth

As human beings appear to be right in the middle of the distance between the largest known dimension of existence and the smallest known component of existence, our planet Earth appears to hold an interestingly similar distinction as to the relationship between the estimated more than many septillion stars and their surrounding planets.

Astrophysicists, biologists, anthropologists, neuroscientists, and learned others, concur that life on earth is possible only because of the presence of a very unique set of conditions. Moreover, the distinction is that we find that there are 11 conditions which exist concurrently and collectively to make life on planet Earth a possibility, the absence of any one of which could disallow life on Earth. Moreover, the presence of all of these same conditions has, so far, not been found on any other body in our Universe.

Interestingly, the National Aeronautics and Space Agency of the United States, and all its foreign counterparts, appear to be vigorously searching and checking to affirm this claim and to further determine if we have any interplanetary cousins waiting to meet us.

Note the precise list of conditions, which together have made life on earth possible:

• Earth's atmosphere is comprised of 77% nitrogen, 21% oxygen, plus smaller amounts of argon, carbon dioxide and water, just right for supporting living things.

• The distance of Earth from our Sun allows the earth to exist with a climate temperature, generally between 0 degrees and 100 degrees Fahrenheit, just right for plant and animal life.

• The magnetic field which surrounds our Earth is just right for shielding the earth against harmful particles from solar winds.

• Water is an essential fluid form, made possible because of the unique temperature range found on planet earth, and, so far, found only on planet Earth, (except, just recently, indications of water in the past have been found on planet Mars).

• The specific presence and location of the Moon makes possible the livable range of ocean tides and the livable range of earth's orbit in relationship to the sun.

• The specific tilt on its axis within a certain range allows Earth's weather to be relatively stable.

• The iron core in the center of our Earth helps protect us from the penetration of the suns radioactive waves.

• The Earth has been around long enough for the beginning forms of life to have evolved into the presence of the human being of today.

• The path of the Earth's orbit around our sun is a shape and distance which allows the ten other essential conditions to exist.

• The Earth is deemed to be of just the right size and orbiting at just the right speed to allow life to be possible.

That these ingredients exist, and that they exist in just the right combination, seems to be more than a coincidence. Does this prove the theory of intelligent design, or does it prove the miracle of evolution, or does this suggest that there is a God which has put all this in place just for the benefit of human beings, or, more credibly, does it suggest that there is simply more to all this than the human mind has so far been able to perceive?

The Pleasures of Life

After examining seven chief characteristics of our existence such as the disparities, realities, dualities, miracles, paradoxes and enigmas, it is time that we look at one more characteristic, this time, something more personal, tangible and positive.

In the busy comings and goings of our lives on planet earth, we should pause to recognize the wealth of pleasures we humans, and only we humans, enjoy while we are here. Yes, there exists an all-too-often burden of pain, suffering, loss and wrong, and these are so very real that we cannot ignore them altogether. Nevertheless, we are somehow uniquely blessed with such an abundance of goodness that the burdens of life can often seem worth the bearing.

There is nothing like the passion of love which seemingly binds together a loving couple, or the sheer joy of welcoming a newly born baby. There is nothing more beautiful than the blossoming of nature at the first warmth of spring, or the contagious feelings which come from the sound of spontaneous laughter. How about the exhilaration felt when observing a mountain waterfall? There is nothing more mesmerizing than the glow of a bright full moon smiling at us from up in the sky. We can't help but notice nature's perfect beauty in a newly blossomed rose, or savor the uplifting feeling of looking across a far-stretching plain of golden wheat.

How about the morning sunrise which celebrates the arrival of the awakening day? There is nothing more gleeful than the laughter of a baby when tickled, and nothing more disturbing than the crying of that same child. And, of course, we must acknowledge the ecstatic sensations of sexual pleasure.

And further still, there is the inner thrill of hearing the gift of song? Does not the sight of a flight of birds across the sky not remind us of the wonder of flying? Do we take time to simply gaze upon the softness of clouds drifting above our reach? Is there not something refreshing and enervating with the passing of a summer rainstorm? Is there not a feeling of rejuvenation which comes from the peace of prayer? How about the warm feeling which comes from the arrival of a special friend? How about the way the sand of the beach tickles our unshorn feet? Is there anything more deeply soul-stirring than the sounds of a cathedral pipe organ? Isn't there something special about the way we awaken from a good night's sleep?

And, too, remember our warm feelings when seated before a crackling fire, the refreshment of a cool drink of water on a hot afternoon, the soothing sensation of the warm water of a shower bath, the reassuring strength which comes from simply holding the hand of another, that simple joy of licking an ice cream cone. And, on and on it goes.

And, the biggest question of all: from where, and just how, does goodness come? Are all these good-feeling sensations simply the workings within our brain? Is this the exclusive domain of the human being? Are these the gifts of an unseen God? Is this the experiential pinnacle of evolution? Will humanity ever really know? No matter the questions, we feel blessed when these sensations are with us.

Reasons for Thankfulness

We have ranged from questions about the void of "nothingness" at the beginning of time, to questions about the "everythingness" of God we ask about today; from revelations about earthly and human centrality, to a reminder of the many unique pleasures enjoyed only by the human species. Every one of these subjects leaves us with questions, the answers to which we will likely never know. Nevertheless, we should be thankful that we are the most advanced component within evolution, occupying the most central position within our known existence, and the principal recipient of things and feelings most pleasurable.

XI.
Paradoxes of Existence

Humanity's Paradise

While humanity's home may not offer the perceived perfection of a Garden of Eden, nor a Heavenly Kingdom, nor a Nirvana, nevertheless, something special seems to be going on which, in so many ways, is tilting evolution's scales in the favor of human beings and the circumstances under which we are living. Just notice:

- We are the most advanced, empowered and complex species as among all living things.
- We dwell on an earth which is singularly positioned to support living things, a habitat so far not found anywhere else in the Universe.

• We are uniquely physically positioned in the very center between the very largest and farthest of creation and the very tiniest and most elusive of creation.

• We appear to have evolved from an earlier level of "consciousness" to that more advanced level we call "conscientiousness".

• We appear to be the only living things endowed with an ability and sensitivity to experience and enjoy a truly broad range of pleasure.

• We appear to be the beneficiaries of evolution's advances, which in turn are offering more and more comforts, benefits and ease.

There are multiple reasons to conclude that humans inhabit a state of total paradise. Such a conclusion, though, is simply not to be.

Devil in the Shadows

I, for one, believe that the notion of there being a part humanlike and part godlike being we call "The Devil" is no more than a human-devised fantasy for explaining away the existence of pain and suffering, loss and emptiness, and evil and harm. Nevertheless, the imagery and notion of the existence of a horrible "Hell" presided over by a terrible "Devil" seems to help us describe what we experience when evil, hurt and wrong come our way.

Clearly there are the pluses of our human existence, but also there are the negatives, together creating a paradox which dominates our existence and defies our understanding. It seems that, whoever we may be and wherever we may be, there is, figuratively speaking, a "devil" lurking in the shadows, just waiting to bring along something bad, like: the death of a loved one, or atrocities of war, or life-taking sickness, or loss of a job, or rejection by a friend, or an economic collapse, or an automobile wreck, or losses and pain of every imaginable kind.

In trying to understand what we could define as "the work of the devil," could it be that we humans simply can't have for ourselves only good and happiness? Could it be that the "bad" is just that other half of the "Dualities of Existence?" Or, could it be that there is so much more to existence than our humanity and our brain can perceive or understand?

It appears that we may have to accept the duality of good and bad, recognize that so much of life is a mystery, acknowledge that there is no real being called a devil, and move on with the most we can understand and the best we can be.

The Devil as depicted on a vintage tarot card.

In My Lifetime

In my lifetime of less than a hundred years, I can count many, many events which appear to be the handiwork of "The Devil"—big world events and smaller personal events. To illustrate my point, though, let me focus on the big ones. Here are what I personally believe to be the big-time, macro-negatives which have come along during the last hundred years:

- The Great Worldwide Depression
- The Japanese Rape of Nanjing
- The Chinese Revolution and Red Guards
- The Great World War II
- The Holocausts of Germany, Soviet Union, and Cambodia
- The Genocides of Armenia and Rwanda
- The Apartheid of South Africa
- The Inhuman Segregation of One Race from Another
- The Korean War
- The Assassinations of Gandhi, King, the Kennedys, Nasser and Rabin
- The Vietnam War
- The Serbian Atrocities Against the Bosnians
- The Gulf War
- The Never-Ending Hostilities between the Israelis and the Palestinians
- The Suicidal Attacks on the New York Trade Center, the Pentagon, and the Attempt on the Capitol
- The Wars of Iraq, Afghanistan and Pakistan
- The Use of Terroristic and Suicidal Tactics by the Young and Disenfranchised
- The Atrocity-Filled Civil Clashes within Iraq, Afghanistan, Pakistan, Syria, Iran, Yemen, Sudan, Somalia, Mali, the Congo, Kenya, Uganda and Nigeria

- The Deadly Tsunamis, Hurricanes, Floods and Tornados Everywhere
- The Atrocity Driven Revolutions in Iran, Syria, Egypt, Yemen, Libya, Lebanon
- The Growing Suicidal Clashes Between Islamist Sunnis and Shias
- The Emergence of the barbaric forces known as "ISIS" and "Boko Haram"
- The East-West Divide within Ukraine
- The Unending Battles for Equality among Races, Genders, Cultures

Human bones in the crematorium of Buchenwald concentration camp after liberation.
U.S. 3rd Army arrived at the camp near Weimer, Germany, on April 11, 1945

If we were to focus only on the negatives and wrongs of society, we would be experiencing a truly hopeless and horrible collapse and destruction of that otherwise privileged species we call "humanity." A factual measure of the wrongs of the world could easily give us reason to just go ahead and consider suicidal death as our only way to personal peace.

While during my lifetime I have known humanity at its worst, I also have witnessed, and even participated in, humanity at its best. There is probably no better measure of some of the goodness which dwells in the human soul than a recounting of those major organizations whose primary mission is to serve others who may be in need. While the factual list runs way beyond our counting, we would do well to pay tribute to and be thankful for the worthy causes listed below. In my lifetime I have known about, even served in, or financially supported, some of the following world organizations. Think about just this partial listing of humanity at its best:

New York, USA—September 27, 2015: President of Ukraine Poroshenko Petro delivers his speech at the UN Sustainable Development Summit in New York

- The United Nations
- The World Bank
- The European Union
- The International Red Cross
- The Centers For Disease Control and Prevention
- The World Health Organization
- Rotary International
- The International Monetary Fund
- The Nobel Prizes for Peace
- The Olympic Games
- World Vision
- CARE
- The Gates Foundation
- The Carter Center
- Center for Civil and Human Rights
- The Martin Luther King Center for Non-Violent Change
- Doctors Without Borders
- World Council of Churches
- UNICEF
- United Way
- National Aeronautics and Space Administration
- The Charitable Outreach of many of the World's Religious Organizations

Throughout my lifetime, I have observed in so many ways that it is often those Christian literalist believers who have given others the greatest measure of caring, sharing, loving, warmth and fervor. These include both my closest personal friends and my most admired leaders. In suggesting the rethinking of religious beliefs and new thinking about scientific revelations as today's strategy for bringing about a more just, equal, caring, sharing, peaceful, and forgiving global society, I also acknowledge the immeasurable goodness of so many of those persons who are firmly anchored in personal religious beliefs which may include differences from my own. These are often valued friends whose work and contributions I have personally known and whose lives have molded mine more than I can measure.

I can't ignore either the courage, beliefs and goodness leadership of those "world saints" who have been the standard bearers for goodness in my lifetime. I am thinking of Mahatma Gandhi, of Dr. Martin Luther King, Jr, of Arch-bishop Desmond Tutu, of Mother Teresa, of President Nelson Mandela, of Dr. Daisaku Ikeda, of His Holiness the Dalai Lama, of President Jimmy Carter, of Ambassador Andrew Young, of Congressman John

Andrew J. Young: Civil Rights Icon, Protestant Minister, U.S. Ambassador to the United Nations, U.S. Congressman, Mayor of Atlanta

Lewis, of Bishop John Shelby Spong, and surely there are many others who have served as the world's champions of goodness.

For me personally, I must add the courage, leadership and loving of Coretta Scott King, the widow who went on to establish events and facilities for the implementation of her husband's insistence on non-violent change as the best way to advance the human condition. My wife and I are privileged to have known this courageous person as a deeply admired personal friend.

When I consider the incredible moral messages coming out of the recent discoveries of "Big Science" and the wonderful moral examples coming out of the lives of these special people, I believe that the time is coming when we can bring all this together in a universal living guide for all the world.

The Greatest Goodness

But then, what do I have to say about Jesus Christ, the one person, who to me, stands at the pinnacle of goodness, the one in all the world to whom more people pay homage than any other, the saint above all saints, the standard bearer above all standards. Be assured that it is the essence of goodness of Christ around which I build and live my own faith. I believe in Jesus Christ as the model for our living and the epitome of our goodness. But, I urge us all to rethink a dependence on the supernatural, a devotion to beliefs not supported by facts, and convictions which are more self-serving than self-giving. It is the self-serving beliefs and the supernatural convictions likely found somewhere in all the world's present day religious faiths that we should rethink. It is beliefs which focus too much on the saving and safety of the individual self and not enough on the well-being of the whole of society. It is this preoccupation with personal self-serving and self-saving, this preoccupation with notions of superiority, this preoccupation with notions of inerrancy, all of which needs rethinking if we are ever to bring about global peace, understanding, and mutuality.

Reasons to Hope

In my view, throughout my life, I have experienced and witnessed far more examples of the positive than I have of the negative. While I may rightfully be counted among the more fortunate, I also have made it my business to be mindful of the negative. Counting all the negatives I can think of, and counting all the positives, there is good reason to earnestly believe that goodness is slowly winning the race over badness, and that it is hope that is the grand force leading us in this direction. There may be multiple paradoxes in our earthly paradise, but there appears to be factual evidence that the painfully slow march of evolutionary goodness is gradually making its way forward with more evidence of goodness than of badness. I see our emerging understanding of Big Science as playing a central role in offering society a recognition of, and an appreciation for, the wonderful things which are happening now and will be happening in the future, what this book defines as the "Dawning of the New Axial Age."

XII.
The Ultimate Message

Axial Happening

I must reiterate that we are living in a time of the beginning of a new axial age. That is, a time in world history when there is occurring in multiple places a seismic shift in human thinking, discovery, invention and belief.

Now, gradually, but surely, we are seeing humans increasingly thinking in terms of the whole world and of all society, rather than just their family, themselves, their community, or their nation. Gradually, but surely, we are discovering information and knowledge relative to the whole Universe, rather than just that of our own Planet. Gradually, but surely, we are discovering information and knowledge about the most basic of matter and energy of which the Universe is comprised, rather than accepting what we see and feel as being all there is to existence. Gradually, but surely, we are inventing devices which interconnect all segments and sects of global society in an instant, rather than leaving us shut inside silos of ignorance and within gates of separation. Gradually, but surely, we are experiencing faith and belief grounded in the real and the tangible of today, rather than being founded on supernatural interpretations of the ancient past.

This seismic shift of so much of our thinking, discovering, inventing and believing is affecting the well-being of all the members of a globalizing society. In the macro sense, we are gradually decreasing poverty, gradually spreading knowledge, gradually sharing governance, gradually improving health and gradually opening minds. In another macro sense, it appears that cultural evolution is underway with big time change, change taking place at an accelerating pace, change moving society in the direction of the good, faster than it is moving society in the direction of the bad. The point here is that this change appears to be leaping forward in a time and at a pace worthy of our attention and support, worthy of recognizing it as an Axial Event.

Connecting the Parts

There are so many pieces and parts to our existence. Think of the Disparities, the Realities, and the Dualities. Remember the Miracles, the Mysteries, the Paradoxes, and the Pleasures, all as discussed earlier. If we are to treat all these as an integral part of a positive new age, there is a big step which society must take. We must bring all this together into specific hopes for the future. We do this by rethinking religion, repurposing science and reimagining society. Our first step is to be open to change and willing to support it. Next comes bringing together these newly defined parts. Then, as Nobel Peace Laureate Mohammed Yunus proposes in his book *Building Social Business*, peace can be brought only by bringing it together "Peace by Piece."

Holistic Perspective

Before adopting a present day, Axial Age perspective for our living, we would do well to put our present existence into perspective.

As to FACTS, which would give truth and reality to our faith for the future, we would do well to first ask: What good is it if we now know when the Universe was formed, or when our planet Earth emerged, or how many stars comprise the Universe, or how fast our Universe is expanding, or whether life will be found on another planet, or if there are multiple Universes, or how our Universe was brought into existence, or how long it took for human beings to emerge, or how many atoms there are on the head of a pin, or how many neurons there are in our brain, or how many neutrinos are flying through the earth, or what exists in a dark hole, or of what is existence made?

As to FAITH, which would give purpose and direction to the discoveries and endeavors in the future, we would do well to first ask: What good is it if we know who, what, where, and whether there is a God, or why and how humanity appears to be the centerpiece of existence, or who qualifies as Divine or Prophet or Saint, or what really happens after our death, or why is there both good and evil, or why is there both pleasure and pain, or is there one correct religious faith, or how can different religious beliefs co-exist, or does prayer really work, or how we should perceive the ancient stories and creeds which are the basis of most religions? And, to those who consider "the chief end of man to be the worship of God," as stated in the Christian Catechism, and as expressed in the tenets of other faiths, I ask, if there is truly an omnipotent and omniscient and infallible God, then what is the need , and what is the role for the human? Do matters really get better for the human if the human worships and prays with fervor, frequency, diligence and subservience to an unseen, immeasurable and un-met God? Has the human being been brought forward through the workings of evolution, or even through the workings of a "grand design," to then exist primarily in the servitude and subservience of a "supreme being," something in existence which is claimed by some to be superior to ourselves?

The answer to all these questions about Facts and about Faith is to follow a simple guide, a first priority, a holistic perspective: Try first to improve the human condition, and strive to do so by employing all that science and religion and society can muster within our human limitations.

Existing Together

The path to a global society of peace, justice, loving, caring, and sharing is through the bringing together into unity science and religion, facts and faith. Until science can embrace the whole, until science can assign purpose to itself, until science recognizes that discovery and invention have a beginning first made of the hopes, beliefs and faith of human beings, then we have not yet established the basis for that partnership between science and religion, a relationship that will enable humanity to discover meaning and purpose, and bring about a newly imagined world for all those beings with whom we share this planet. Then, until religion can comprehend the whole, until religion can embrace truth to itself, until religion recognizes that faith and belief must first be built on facts and realities which exist beyond humanity, then we have not yet established the basis for that partnership between science and religion, the basis for that partnership that will enable humanity to venture into methods and means for establishing and sustaining goodness as the heart of our existence.

The reality is that a meaningful, honest and truly useful religious faith needs facts for its justification as much as does any form of science. Likewise, a purposeful and advancing science needs hope, belief and faith to undergird its search for purpose and usefulness. In that sense, neither science nor religion can serve the common good by itself. Each needs the other if it is to serve humanity. What then if we set out to bring the two together?

Defining "God"

Most of us are mindful of the fact that there exists a very broad range of thinking about that entity so many call "God." There is the conviction of the Atheists that there is no such thing as God. There is the conviction of the Agnostics that understanding or confirming the existence of God is beyond our human ability to perceive and beyond science to confirm. There is the conviction of the Pantheist that God is the sum of all being, the aggregate of all existence. There is the conviction of the Panentheist that God infuses all being, but also exists separately, beyond all being as the creator and source of all being. There is the conviction of the Theist that God is a divine entity above and beyond all else, while still sharing characteristics with human beings. There is the conviction of the Humanist that the human is the centerpiece of all existence and that whatever characteristic we may or may not attribute to a God, it is there for the creating and sustaining of the human being. And for many more, their God is somewhere in between or slightly different from any of the above.

In considering that range of ways of understanding God, what is most important for our present purpose and need is to re-examine closely those beliefs and faiths which most defy an adaptation to a universal faith, that universal set of beliefs needed for the sake of and survival of humankind.

The greatest failing of all present-day major faiths is the human propensity to imbue the "divine" with human characteristics. Such attribution is known by a word, anthropomorphic, "the ascribing of human form and attributes to a being not human."

As previously discussed, painted on the ceiling of the Sistine Chapel of the Vatican in Rome is Michelangelo's most famous depiction of a grandfatherly-like God figure in an act of first creation, reaching across a colorful heaven to touch and bring into being the first human male form. For far too long, so much of humanity has been groomed to believe that all existence, and all humanity with it, was created in an instant by that grandfatherly depiction of the image of God. Furthermore, we were told that it all began out there somewhere in a place called The Garden of Eden. Such imagery is reinforced by the Biblical claim that we humans are created in the image of God, not only suggesting that we humans are blessed to look like God, but also implying that God looks just like us.

Do we not recognize that if God looks like us, then we are defining a terribly weak, impotent, faulty being who could not possibly have the power and knowledge to create all existence? What a disappointment. If we humans have God-like characteristics, then where is our infallible knowledge about everything? Where is our boundless power to create everything. Where is our overriding capacity to correct the ills of our existence? Out of this anthropo-morphic imagery, this concept of divinity, arises the belief in an all-powerful, all-knowing being "out there somewhere"—a being who created existence, directs its destiny, establishes its rules, selects its favorites, and demands its obedience.

Nearly half of present-day humanity rejects the notion of biological evolu-tion, and more than half believe in a never-seen, never-proven, always right, human-like being whom they call God, or Allah, or Yahweh or other super-human or supernatural names, each signifying a magical and majestic being who resides in a space called a heavenly kingdom, or the like. Lacking any tangible proof of this supreme being, surely this is a concept housed in the human mind, a concept born out of the human hunger for something bigger, stronger, and better than ourselves.

When I observe the recent discoveries of natural science, or what I think of as Big Science, and consider its scientific proofs that existence is a supremely boundless and supremely complicated something, I find neither evidence nor justification for a separate, all-controlling being existing separately from us in an infinitely distant and paradise-like place called Heaven. That definition and concept suggests to me that we need to get past that unfounded and de-meaning idea and move our lives into the realities of the present.

Reality of Faith

In striving to understand the existence or nature or description of God, I also acknowledge that there is likely to be so very much that my human mind does not know and cannot perceive. Understanding the existence of a god, or the nonexistence thereof, is the preeminent question which tops humanity's list of what still remains unknown.

For me, I, first, acknowledge that I only know what I know as a human being. Some of what I know may be incorrect or obsolete or incomplete, but this carefully considered being finds that the most intelligent definition is a belief, not a fact, a belief that God may well be our term for the sum of all existence, a definition for all existence. If that be true, then it would happily follow that I, as a human, am also part of God, and that God is also a part of me.

Reality of Me

Knowing the limitations of my brain to perceive, and knowing the unlimited bounds of all existence, I still find a need for a definition of my being and a guide for my living. In striving to follow factual knowledge, in confronting immovable beliefs, and in acknowledging human disparities, I find multiple reasons to lay out a guide, a "universal code of living" for all humankind, a "foundation for faith" for the new Axial Age now upon us.

Before we attempt to adopt a strategy and establish guidance for living in the new Axial Age, I'll share with you a personal story. It is a story which offers an essential perspective and a foundation for what we may wish to become "The Next Axial Age."

Laughing at the Devil

Earlier that morning, our host had taken us to the thatched-roof, mud-walled, dirt-floor home of a typical family. We were north of the town of Kisumu, in Africa's Republic of Kenya, almost within sight of Africa's great Lake Victoria.

First, a tour of the yard. Circling the hut-like home were strange mounds of freshly turned earth, and we were puzzled by this sight. Invited inside the humble home, we were welcomed with the warm greeting of a proud-faced, but ill-dressed couple. Sitting on our hosts' tree stump chairs, we learned that we were in the heartland of the world's worst AIDS afflicted country. To prove it, we were told that it was members of their extended family, all related, who were the occupants of those crudely mounded earthen graves surrounding this crudely structured dwelling. Lying in each of those grave sites was the body of a victim of AIDS.

But then a little later, there was to be: "that woman."

My Rotary Club colleague and I had traveled from our Atlanta home to this place, a place I quickly saw and felt to be the world's purest example of what poverty, death and disease were all about, a place urgently needing all the help we could bring. As for us, our mission was to visit and explore what our 500 member club could do to assist with the survival of these beyond-belief, most-distressed fellow human beings.

Next, we were seated on the front row, to be treated as the honored guests in a simply built, open-sided community gathering place, a gift of a Lutheran Church from I know not where. On the make-shift stage, about 20 of our female entertainers were ready to pour out their hearts to offer their best to these visiting would-be saviors of their humble society. While waiting for the opening act, our host/guide whispered that the dozens of little boys and girls who stood encircling the seated audience in this simple place were each the orphans of parents who had died of AIDS.

Disturbed enough by that whispered revelation, a feeling of shock and grief pulsed through our heart-sick bodies when we were then told that most of these young orphaned children were born HIV Positive, the real message being that before long, they, too, would likely be the occupants of another one of those freshly turned graves.

But, then, there was yet to come: "that woman!"

The "entertainment" was a singing and dancing skit, first about the scourge of AIDS, then about living in mutual encouragement of each to the other. While the dancers jumped their highest, and raised their voices to their limit, what was going on quietly inside me was a melted heart and shaken mind. Tears rolled down my red-flushed cheeks. This already "luckiest guy in the world" was being treated with a never felt before passion and love and respect by some of the unluckiest people with whom I share this earth.

And then, there was "that woman!"

She was on the front row of the stage, in the center. If there was clapping, hers was the loudest. If there was dancing, her feet and arms swung more widely than those of any other. If there was singing, her voice was raised louder than all the rest. If there were smiles, hers was the widest. This, our most earnest entertainer, was of stunted height, of misshapen body, of twisted arms, of bowed legs, and, bluntly, of a gnarled, and ugly face.

Of course, she too, was the partner and mother of lost lives lying in those graves we had seen earlier, while she was herself HIV Positive. As if that were not enough, she had chosen to be the adopted mother of some of those sad-eyed and orphaned children quietly standing around us. Never before had I ever witnessed, or even imagined, a human who faced life with the courage and grace of "that woman"—an incredible person who was selflessly giving her all to the life she had been dealt, and sharing her best with those two awestruck healthy men from another part of the world, her royal guests for the day.

My friend and I left, not feeling like talking, but holding in our silent selves a prayer for those good and struggling people, and a promise to ourselves that we would bring back help. For me, I realized that, if our humble and hurting hosts could still view life with such fervent faith and hope, I would be returning home with an exemplary lesson which would guide the rest of my life.

XIII.
The Axial Mandate

The Sum of the Parts

We began with the suggestion that there are many signs that we may be entering our own new Axial Age, a newness brought about by the coming together of four very significant, perspective-changing experiences:

- The accelerating pace of Big Science revelations about all existence, including our human existence.
- The increasing empowerment of individuals to think and act more for themselves.
- The broadening of knowledge about other people and societies around the world.
- The shifting from dependence on belief in the supernatural to the independence offered by belief in the fact-based real.

With what appears to be a positive and encouraging new perspective about our existence, there is good reason for us to take a closer look at what may be happening, or better yet, at what we can make happen.

Rethinking Religion:
From Heavenly God to Sacred Human

At the beginning of this book, I presented the claim of philosopher Karl Jaspers that there seemed to be a First Axial Age, a time during the years from 800 BCE to 200 BCE. It was a time in which the thinking human emerged, a time of the appearance of Aristotle and Socrates, to the laying the foundation for the later time of Jesus and Mohammed. Humanity generally seemed to move away from a dependence on and fear of nature's gods of sun, moon, rain and fertility to a more human-minded time, beginning with wise human philosophers to newly forming religions of Greek Philosophies, Jainism, Ahimsa, Buddhism, Hinduism, Taoism, Confucianism, Zoroastrianism and Judaism.

In most of these religious societies, life and belief were formed around an all-powerful and unseen, but much felt God figure thought to reside in a separate and distant place, most often thought to be way up in the sky. There was even the belief that this incredible God appointed earthly representatives. This idea added still another dimension to omnipotence and omniscience.

Further thought suggests to me that it would now be correct to believe that this original Axial Age has been followed by a Second Axial Age, an age which generally has run from around 200 CE to around 1900 CE, a time of invasive conquering, such as that of the Barbarians, Huns, Visigoths and Genghis Kahn, and later the Ottomans and Christians; a time of global population relocation, such as the European populating of the Americas; a time of revolution enforced political change, such as the Islamic Ottomans, the Christian Crusades, the French Revolution and the American Revolution; a time of geographical exploration, such as that led by Columbus, da Gama and Desoto, to Stanley and Livingston, to Lewis and Clark; a time of colonization of the weaker, such as that of the European takeover of all or parts of Africa, India, the Americas and East Asia; a time of enslavement of the defenseless, such as the African slave trade to the Americas; a time of emergence of organized religion, such as that of Christianity and Islam; a time of reordering existence, such as the claims of Isaac Newton, Galileo Galileo and Charles Darwin; a time of brute enforcement by centralized power, such as that of the Kings, Emperors and Czars dominating nations around the world. In summary, this Second Axial Age proved to be an historic era of ceaseless military, governmental, societal and religious churning and upheaval.

There are convincing reasons to believe that we are now entering a Third Axial Age. As to the dating of our new Third Axial Age, I find numerous examples of significant change into something quite new and different, beginning with that time of the inventive wizardry of automobiles, airplanes and rockets, of electricity, lighting and motors, of medicines, inoculations and sanitation, of radios, telephones, television and computers, of peace-winning World Wars, United Nations and diplomacy, of astrophysics, quantum mechanics and neuroscience. As a generalization, I see this accelerating pace of societal change beginning around 1900 CE, signaling signs of the emergence of a Third Axial Age.

I see the First Axial Age, that time in history observed by philosopher Karl Jaspers, as the beginning of the *Thinking Human* and the birthing of philosophical and religious thought. I see the Second Axial Age as that long time in history of the *Societal Human* and the evolving of humanity into organized cultures, beliefs and governments. I see the now arriving Third Axial Age as the time of a newly empowered *Global Human* and the gradual evolving of humanity into increasingly universal practices, beliefs, governance and concerns. As to what is happening today, I see us eventually, but surely moving away from the "Me-First-Ness" of populism and nationalism to a gradually emerging mixing, connecting, combining, interdependent, world society, offering humanity the real opportunity for more consensus building and less independence demanding.

The notion of there now taking place a "Dawning of the Age of Aquarius" suggests to me the possibility of the arrival of the most exciting, promising, encouraging, positive and civil happening in world history, perhaps, and hopefully, the dawning of an "axial age of togetherness." Perhaps the most dramatic evidence of this historic arrival of "togetherness" is the 2015 uniting together of the major world powers, and history's one-time enemies, the United States, China, India, Japan, Russia, Britain, France, Italy, Germany, to collectively deal with two global issues. One of these is the multi-national shuttle diplomacy effort to rid the world of the barbaric acts and apocalyptic-thinking of the Islamic State. The other is the coming together of 190 nations at a conference in Paris to adopt a global plan for reducing the very real earth-flooding consequences of Climate Change. That these nations came together in the same year to address two of Earth's most destructive issues offers proof that historic events have taken place, that "globalization" has arrived, and a Third Axial Age has begun.

While hoping for such a reimagined world, I suggest that such will not happen until and unless religious and cultural beliefs are first loosened from their fear and ignorance-based cultures built on ancient convictions of an omnipotent and omniscient anthropomorphic God promising us a better life after this one. This also means that as long as the tradition-based faiths of today fail to open themselves and adjust themselves to the science-based revelations of today, then fighting, injustice, envy and disparities will continue to rule the day.

The encouraging news is that half the world is beginning to view matters of change more openly, that Abrahamic faiths are witnessing perspectives quite new, that science is crowding us with information so thought-provoking, that people are now mixing with others quite different, and that change is exploding with dramatic newness. Such realities lead me to the hope that there is good news standing before us, good news if we will only open up our minds and let this new goodness enter into our lives.

Repurposing Science:
From Abstract Discovery to Purposeful Revelation

When we think of "science" today, we too often are drawn to think of an on-going and ever-running array of inventions, conveniences, and devices serving our lives with greater comfort, safety and power. However, the real gift of a Repurposed Science is a purposeful guide for our living, not just an abstract device for our getting. The real gift of our time is found, not in the material and the useful and the tangible, but instead in the over-arching moral guidance born out of the recent revelations of astrophysics, nanophysics, particle physics and neuroscience. Consider the following:

When we consider the vastness of the Universe, the speed of its changing, the complexity of its contents, and the finitude of its survival, there is good reason for us to be struck with an overwhelming sense of awe and humility. There is no place for humans to feel superior to matters around them or in relations between them. The moral message of the revelations of astrophysics and quantum mechanics is that we should live our lives in awe of our existence and with humility towards those with whom we share it.

When we consider the composition of existence, we would do well to remember that all existence is comprised of the same three components: upquarks, down-quarks and electrons. The moral message of the revelations of particle physics is that, since all that exists is of kindred composition, we should live our lives with profound reverence towards all which surrounds us and with profound respect towards those with whom we live.

When we consider the limits of our knowing and the boundaries of our brains, there is good reason for us to recognize that we do not even begin to know how much we do not know. There is no justification for us to assume that we know all which matters when science offers absolute evidence that we can know only that which our brain has been able to perceive. The moral message of the revelations of neuroscience is that we should live our lives with an eagerness to learn all that we can and a recognition that there is so very much more that we do not know.

When we observe the centrality of the human being, the dualities of our existence and the benefits of all our privileges, there is good reason for us to recognize that the human being appears to be the most unique, complex and advanced of all living matter within our Universe. In this way it appears that the human, as the recipient and guardian of existence, carries a responsibility towards all that lies before us. The moral message of the revelations of all of Big Science is that we should live our lives with a sense of responsibility and thanksgiving towards all the good which surrounds our personal existence.

Reimagined Society:
From Fear-ful and Self-ful to Fear-less and Self-less

A central factor or foundation of the Third Axial Age is the growing empowerment of the individual, as compared to history's domination by the few. Three significant empowerment events are taking place simultaneously. Collectively they offer society real hope for improving the human condition and they offer society good reason to reimagine what can still be.

First, there is the real and measurable decline and displacement of authoritarian governments, and with it the elimination of power and control by the few. For example, in less than half a century, the dictatorship governments of most of South and Central America have been replaced by individually empowered democracies. Although the change is not yet absolute, similar progress is being made throughout most of Europe and Africa.

Second, there are the real and measurable advances of science and technology, with such advances yielding increasing empowerment of the individual. For example, instantaneous communication devices available around the world are empowering individuals with instantaneous information and broadened knowledge. Ever quicker travel to places around the world are empowering individuals with increasing experiences and broadened exposure. Together, advances in communication and travel are empowering the individual with capabilities and freedoms to more wisely think and act for himself or herself as never before.

Third, with these newfound capabilities, the individual is more informed as to the disparities and needs of society. Then, the individual is more free and more available to do something about them. The more free and better informed the individual, the more likely may be the individual to act on the needs of his or her time. For example, observe that it is the democratic freedoms and the technological advances of the United States which have led its society to operate the widest range and largest number of non-profit and charitable organizations in the world. It seems that freedom and knowledge for ourselves have been the leading factor for our greater caring for others.

While the previous listing of the Disparities of Our Existence reveal widening gaps of knowledge and wealth and resources for many, humanity is nevertheless advancing with powers to think and change and act as never before.

If this individualized empowerment is lifted along with the forces of evolution, and if this individualized empowerment is welcomed along with openness towards change, and if this individualized empowerment is encouraged along with acceptance of universalization, then we can have a reimagined society wrapped in a good measure of selfless love.

Such reimagining of the society in which we live, and understanding of the Earth on which we live, will not come as easily as our capacity will allow, but will come as surely as our will may allow. If the two previously defined Axial Ages are measured by centuries of time, we have no right to expect instantaneous change in the third. But if we have found that the existence of the Universe, and all within it, is moving faster and farther sooner, then the prospect for a Reimagined World Society is worthy of calling on each generation to give its very best towards the speeding along of its fulfillment.

The Axial Guide:
Hope Melting Into Reality

It seems that history has forever offered declarations, commandments, creeds, prayers and proclamations for the guidance of the society of its time. Think of the Code of Hammurabi, the Ten Commandments, the Beatitudes, the Lord's Prayer, the Nicene Creed, the Apostles Creed, the Magna Charta, the American Declaration of Independence, the Constitution of the United States of America, the Emancipation Proclamation, the United Nations Declaration of Human Rights, even the recent Charter for Compassion offered by theologian Karen Armstrong.

Along the path leading to a time of hope, goodness and happiness for all, there is good reason for us to ask: what shall be our guideposts for arriving at our newfound destiny? I propose that there be a new universal code for our living, a Creed for the New Axial Age.

XIV.

A UNIVERSAL CREED
FOR THE NEW AXIAL AGE

Live in Awe of All Existence

Recognize the Limits of Our Knowing

Treat All Life as Precious

Honor the Kinship of Everything

Care for and about Each and Every Human

Rid the Wrongs of Human Disparities

Give Away Our Love Unconditionally

*Accept That All That Is of Matter Is Born, Lives,
then Surely Dies*

*Understand That No One Has Anything to do with
the Circumstances into Which He or She is Born*

XV.
OUR AXIAL WORLD DEFINED

Waking

While no small measure of imagination and belief are required, there are abundant reasons to think that the Third Axial Age is already walking through the door into our lives today, leaving us with the imperative that we start now to embrace its challenges for change.

In 1890, the Nobel-winning British poet Rudyard Kipling wrote his memorable poem, "On the Road to Mandalay," a beautifully described experience made all the more famous by the world renowned singer, Paul Robeson. I remember this, because this was my father's favorite, and it became an indelible memory for me. Borrowing from this most descriptive and beautiful poem, I find the phrase: "On the road to Mandalay, Where the flying fishes play, An' the dawn comes up like thunder ..." as an apt description of the dawning of our new axial age. Like "the flying fishes play," while the world is preoccupied with the playing out of daily living, there is the distant sound of a new age "coming up like thunder," a signal of exciting change coming our way, change not to be taken lightly. It is the Dawning of a New Axial Age.

Change

The change or evolution coming our way will be carried in, not by the entrenched institutions of religion, or education, or government. Change will come from the newly empowered individual. With the arrival of all the devices which have aided the advancement of humanity, from the arrival

Smartphone users on a Hong Kong subway.

of the wheel, to sailing ships, to electric lighting, on to farm tractors, family automobiles, jet airplanes, interstate highways, miracle medicines, telephones, televisions, computers, and whatever else you wish to name, there is no comparison to the speed of arrival of that latest phenomenon, that miracle generating smartphone and its social networking, offering the opportunities for instantaneous connection on every imaginable subject with other humans living all around the world.

Acceptance and use of instant connection by way of social networks and hand held instruments is the fastest growing humanity contrived phenomenon in the history of the world. In only a decade or two, half the world has come to have an email account. Purchase of the newest smartphones is growing at a compounding rate of 60% per two years. The pace of the emergence of social networking is accurately and amusingly defined in an article by Nelyeska Hernandez in *Eelanmedia*, paraphrased here: "People are waking up in the morning to first checking personal messages on Facebook and Twitter before they use the toilet, brush their teeth, have a shower, or drink their coffee."

All of a sudden lives are being connected by a growing proliferation of Microsoft, Wi-Fi, Yahoo, Gmail, Facebook, Twitter, Google, Linkedin, Instagram, YouTube, Snapchat, Alibaba, and who knows how many more new forms of communication and connection are on their way. Each and every one of these is empowering and changing both individual lives and entire societies at a pace never known before. Be assured, the Dawning of the New Axial Age, is galloping in on the back of science, and at a pace beyond our measure.

Humanity

As the individual is being increasingly empowered as an individual, there is no longer the propensity to "follow the crowd." In its place is the encouragement to "think for one's self." Therefore, there is no longer the incentive to anchor one's life in "safe harbors" and inside "gated communities." The yearning, the enticement, the tendency is for one to open one's life to venture out on to the high seas of exploration and discovery.

In the past, the United States has been called a "melting pot" of the world. In the new axial age, the whole world gradually will be experiencing an acceptance and easing of the coming together of those who, in various ways, may be different one from the other. Slowly, but surely, the richness of diversity will replace the safety of sameness. While slowed and delayed by the set characteristics of the culture into which they were born, individuals will be increasingly enticed to venture outside their inheritance, learn more about the rest of the world, and be more comfortable with the mixture of customs and thoughts of the people who inhabit it.

For thousands of years society has been beholden to the notion that the supernatural and the sacred are one and the same. That which is beyond our complete knowing and confirming has been set apart and defined as sacred, meaning: to be treated with deference and subservience, oftentimes to the point of abject fear. I find so many of religions' demands to worship God, to confess our sins and to reject non-believers to be examples of religion's claims of being sacred and superior. For the three thousand years comprising the First and Second Axial Ages, most of humankind has viewed its place as subservient to an omniscient and omnipotent power reigning from above all viewable existence.

Gradually and slowly, humanity, in the form of its newly empowered individuals, is establishing itself as freed from subservience and deference to the point that we seem to be beginning to see the human acting as the "sacred" and their notion of God as simply the "unknown." Chapter IX offers more on this notion with its commentaries on "The Simultaneity of Nature," "The Centrality of Humanity," and "The Uniqueness of Earth." For an increasing number of the younger generations the new empowerment brought on by science and technology seems to be offering them an independence and confidence which frees them from a need for, and dependence on, matters of religious faith. Whether or not humans deserve this newly proclaimed throne of "above-ness," we are seeing evidence that in the new Axial Age we will experience quite a difference in any understanding or practice of the relationship between God and the Human.

Knowledge

Until rather recent times, society has relied on the formally organized schools, colleges and universities as the primary source of gaining knowledge. While these essential institutions will likely remain in the forefront, the measure of the quantity of knowledge being dispensed is fast being challenged by the global connecting, all-knowing device, the internet. Here the thoroughness of word-knowledge dispensed via books is fast giving way to the quickness of image-knowledge dispensed via screens.

There is another phenomenon affecting our knowledge and our learning. For so long, learning has come by way of classrooms of students listening to the oral presentations of their teachers. Fast succeeding this method of learning is that made possible by the personal computer. Instead of receiving doses of information taught by a teacher in a classroom, students are given assignments to do their own learning by way of using their computer as a gateway to a tremendous amount of information stored offsite "in the cloud."

As discussed in the "Disparities" listed in Chapter V, knowledge and information are perhaps the fastest growing component of our existence, or more certainly of our societal living. Strangely and sadly, we are finding that this very same growth in knowledge and information is leading to an exponential spreading of the divide between the haves and the have-nots, and between the informed and the ignorant.

We would do well to follow closely the path of "knowledge gained." There is reason to wonder whether, in the new Axial Age, science will come to the rescue of "needed knowing" or will it only accelerate the societal divide of "knowledge versus ignorance?"

Timing

Before we are too soon carried away with the excitement and promise of the currently advancing New Axial Age, the reality of time trips up our hope, excitement and motivation. While it is acceptable and consoling to think of the words of Rudyard Kipling and his words of "dawn comes up like thunder from 'cross the bay,'" and while it is also only human that we hope and think that a great, positive new age is rushing in, we will likely need to recheck our watches and calendars.

It will not be easy for Western Civilization to let go of its precious traditions, where Americans proclaim on their currency: "In God We Trust", and in their Pledge of Allegiance: "One Nation Under God," and in their singing: "God Bless America". All this seems to be an inheritance from America's English forbears who ask that "God Save The Queen." These are long standing and deep rooted government endorsed proclamations which serve as an anchor of faith for its well-intended citizens. In turn, all this is reinforced by the religious celebrations of Rosh Hashanah and Yom Kippor, Christmas and Easter, Ramadan and Mecca.

Recognizing these entrenched traditions in part of the world, we must also recognize that there is another part of the world which thinks differently. The nation of China alone is comprised of more citizens than Western Europe and the Western Hemisphere put together. In turn, China is now being recognized as the fast arising dominant nation, culture and economy of the world. With this in mind, we would do well to recognize that China already disallows allegiance to and dependence on religious faith. Does this not suggest that a major part of the world has already been "rethinking religion?"

The reality is that the "dawning of our age of Aquarius," and the "arrival of our new Axial Age," even in an exponentially evolving existence, is foreordained to be measured not in months or years, but in decades and generations. Such a pace will demand of us, cruel-hearted patience on one hand,

and undaunted determination on the other. Perhaps we can be consoled by the hope that the changes to come will be for the betterment of humankind.

Progress

Be assured that the positive notions proposed in this book are not oblivious to the vicissitudes of existence, nor are the hopes of its author naïve. The long reach of evolution is not a straight line, nor do its workings offer a smooth path. The arriving of a Third Axial Age does not mean that humanity will arrive instantaneously at a state of blissful nirvana. Existence and evolution just don't work that way. What is proposed is that an increasing number of people on this Earth are being increasingly equipped with new powers and knowledge with which to make life increasingly better for more people and to do so more quickly.

A look at history reminds us that progress seems to come in irregular and uneven steps. Sometimes there is regression before progression, and then sometimes progress is followed by setbacks. Observe some examples. The world had to be led by kings and emperors and dictators before elected presidents and prime ministers arrived. Barbaric invasions prevailed before the enlightenment of the Renaissance era arrived. European Colonialism dominated the world before independent democracies could follow. Slavery prevailed before civil rights elections eliminated it. The horrors of holocausts took place before exemplary societies could emerge. Now, in these times, the atrocities of Islamic terrorism and religious divisiveness must be dealt with before peace can be with us.

All of this is to say that we will hope and strive for the now arriving Third Axial Age to be history's greatest time of justice, peace, equality and compassion. As in the past, these better times inevitably will have their share of the negatives of cultural upheaval, civil violence and natural disaster. Nevertheless, there is reason to believe that these negatives will be counterbalanced with a greater measure of positives.

As sociologist Michael Shermer would remind us, it is the net gain of progress and goodness which counts. And, I would remind us, it is the nature and extent of this net gain which gives reason for us to be excited and encouraged about this Third Axial Age into which we are now entering.

Courage

In most realms of human endeavor, the word "courage" is an attribute best illustrated by the courage of a soldier under intense fire in wartime combat, an occasion of extreme danger, and an occasion of uncertain survival. Surely this is an appropriate description of "courage" in the midst of the war, terror, and hate which are all too prevalent around our ever-dissonant world of today. After the death and destruction of two World Wars, one would think that humanity has had enough, that courage has done its job.

But the wrong goes on. We cannot look away from the dozens of "little wars" surrounding and dominating the world today. Pope Francis of the world-spanning Catholic Church has defined today's events as "A Third World War…In Pieces." Russia's Nobel Peace Prize winning Mikhail Gorbachev fears that global-scaled war is returning. While we should never forget the courage of the innocent humans caught in the midst of wars and turmoil, be they local or global in scale, there is a mandate for humanity today to offer to this restless world still another form of courage.

If society today is to receive the waiting blessings of a New Axial Age, it must demonstrate another degree of courage worthy of the blessing. As much as anything or any form or any gift, society needs the courage of choruses of loud voices proclaiming the opportunities before us and denouncing the societal and religious barriers which stand in their way. Remember, we are dealing with a present day society so deeply caught up in backward pushing disparities, that the courage of loudly calling voices speaking for the goodness of the new is the overriding need for our time.

If we are witnessing the increasing empowerment of the individual, then the courageously exclaiming voices must come, not just from the collective voice of the established, but more likely out of the restless vigor of the newly empowered young. What the world needs so desperately is for the newly empowered young to demonstrate personal courage supported by their unlimited new knowledge. It is the newly empowered and informed individuals upon whom the mantle of change is now placed.

It is propitious and relevant that we borrow from the wisdom of Thomas Arnold, Headmaster of England's Rugby School, who, disturbed about the troubles of his native country in 1835, declared in a letter to friend Reverend F.C. Blackstone: "[…] I seem to find it more and more hopeless to get men to think and inquire freely and fairly after they have once taken their side in life. The only hope is with the young."

In spite of Thomas Arnold's admonition, society already overwhelms our young with the burdens of fighting our wars, getting an education, finding a job, establishing a home, and rearing a family. Strangely, it is these burdens which prompt the young to be discontent with the present, to be open to change, and ready to seize upon that which seems better.

Conversion

In my lifetime, I have known many who claim to have been "born again." As stated before, I have found these persons, many of whom are valued friends, to be among the most genuinely caring and sharing among all whom I have known. While I strongly disagree with the self-saving theology which guides so many of them, their devotion is to be emulated.

I must first admit to a sense of sadness and concern for those whose lives are built around the claim that their subservience to a long ago Jesus Christ, or to Mohammed, or to Buddha, or to a distant heavenly being will carry them into an eternal and paradise life to be enjoyed forever in the company of their god and their kin. We generally define this heavenly granted conversion as "Being Born Again." While I wish the best for my "born again" friends, I would wish for them even more the enlightenment of reality.

For me, I have already found a different form of being "Born Again." It was something which pulled at my very being when I attended the service at Ebenezer Church. It showed up again when I witnessed the story of The Universe. It tore at my very soul when I visited the settlement near Kisumu. In fact, it has haunted me for most of my life. It has not been the "saving of my soul" which has counted most for me, it has been the "serving of others" which has come to be the very essence of my being.

All of this is to say that, in the realm of the New Axial Age, I feel certain that "Born Again" will not be the measure of the "chosen" or of the "goodest," but "Born Again" will be the measure of a sharing and caring life. It is only this version of "Born Again" which will bring to this earth peace, justice, equality, sharing, caring and loving, the defining marks of the Third Axial Age.

Daring

While I have already addressed the topic of Courage, I close with that characteristic which best defines whether the New Axial Age will bring us gain or loss, joy or struggle, peace or terror. I detailed earlier the Mystery of Duality, that is, the strange, mysterious, unwelcome characteristic of all existence: the presence of evil as seemingly the price of good, the presence of suffering as seemingly the price of blessings. While we should not go lightly into the concerns about these anomalies, and we should not look upon them with easy concern, there may be another lesson which awaits our discovery.

There is good reason for us to pay close attention to the lesson from Ebenezer Church, to the message from "We Shall Overcome, Some Day." While the wait for freedom and equality for those Americans of darker color lasted a few hundred years, once the enslaving barrier was put down and the gates of freedom were opened, it has been a fewer short years that these same gates of inequity for persons of a different gender, and these same gates, which have locked in persons of different sexual orientation, are now more quickly opening around the world.

Along the way, change would not have taken place, hope would not have been answered, equality would not have been allowed, and love would not have prevailed had it not been for the initial daring of a smaller number of courageous people. In this hope-filled new age there is again a call for that daring. Let each of us be among the loud-voiced, the courageous, the daring, then the goodness of the new Axial Age will come the way of the world.

Then, just maybe, that goodness will come sooner than we may think and, just possibly, it will be greater than we could ever imagine.

Vision

As we move forward on our journey to do all we can to improve the human condition; as we strive to bring together the inspiring lessons from the discoveries of Big Science and the core of goodness in the world's religions; as we dream of a world of peace, compassion and togetherness; and as we attempt to lead the way into a New Axial Age; we may find both comfort and encouragement in the wisdom of famed American astronomer and astrophysicist, Carl Sagan. Consider now the prophetic statement written by Sagan only a few decades ago:

"A religion old or new,
that stressed the magnificence of the Universe
as revealed by modern science,
might be able to draw forth reserves
of reverence and awe
hardly tapped by conventional faiths.
Sooner or later, such a religion will emerge."

Carl Sagan
Pale Blue Dot
1994

Works Cited or Referenced

Armstrong, Karen. *Fields of Blood.* New York: Alfred A. Knopf, 2014.

Brownowski, Jacob. *The Ascent of Man.* New York: Random House, 2011.

Carter, Jimmy. *A Call To Action.* New York: Simon and Schuster Inc., 2014.

Cosmos: A Personal Voyage. Dir. Carl Sagan. Carl Sagan Productions, 1980. DVD.

Cosmos: A Spacetime Odyssey. Dir. Ann Druyan and Neil DeGrasse Tyson. Twentieth Century Fox, 2013. DVD.

"Infographic: The Global Education Crisis." BuildOn.org, 25 Sept. 2014. Web. 27 June 2015. <http://www.buildon.org/2014/09/infographic-the-global-education-crisis/>.

Jaspers, Karl. *The Origin and Goal of History.* English Translation. New Haven: Yale University Press, 1968.

Kotkin, Stephen. "The Communist Century." The Wall Street Journal, 2 Nov. 2017.

Kristov, Nicholas. "Why 2017 Was the Best Year in Human History." The New York Times, 6 Jan. 2018, www.nytimes.com/2018/01/06/opinion/sunday/2017-progress-illiteracy-poverty.html.

The Dalai Lama. *The Universe In A Single Atom.* New York: Random House Inc., 2005.

Naisbitt, John. *Megatrends 2000.* New York: William Morrow & Co.,1990.

Primack, Joel. *The View from the Center of the Universe.* New York: Riverhead Books, 2006.

Sagan, Carl. *COSMOS.* New York: Ballantine Books, 1980. Print.

—. *Pale Blue Dot.* New York: Ballantine Books,1994. Print.

Shermer, Michael. *The Moral Arc.* New York: Henry Holt and Company, 2015.

Wikipedia contributors. "List of wars and anthropogenic disasters by death toll." Wikipedia, The Free Encyclopedia. Wikipedia, The Free Encyclopedia, 13 Jan. 2018. Web. 14 Jan. 2018.

World Bank. *World Development Indicators 2008*. World Bank, Apr. 2008. Web. 12 Nov. 2016.
<http://elibrary.worldbank.org/doi/book/10.1596/978-0-8213-7386-6>.

World Bank. *World Development Indicators 2016*. World Bank, Apr. 2016. Web. 12 Nov. 2016.
<http://elibrary.worldbank.org/doi/book/10.1596/978-1-4648-0683-4>.

GENERAL BIBLIOGRAPHY

RELIGION

Armstrong, Karen. *A History Of God.* London: Gramercy Press, 1994.

—. *Fields of Blood.* Alfred A. Knopf, 2014.

—. *Twelve Steps To A Compassionate Life.* New York: Random House, 2011.

Borg, Marcus J. *The Heart of Christianity.* New York: Harper Collins, 2003.

—. *Reading the Bible Again for the First Time.* New York: Harper Collins, 2001.

Coogan, Michael. *World Religions.* New York: Metro Books, 2012.

Crossman, John Dominic. *Excavating Jesus.* New York: Harper Collins, 2001.

—. *God and Empire.* New York: Harper Collins, 2007.

Eckel, Malcolm David. *Great World Religions--Buddhism (The Great Courses).* Chantilly: The Teaching Company, 2003.

Esposito, John L. *Great World Religions--Islam (The Great Courses).* Chantilly: The Teaching Company, 2003.

Gafni, Isaiah. *Great World Religions--Judaism (The Great Courses).* Chantilly: The Teaching Company, 2003.

Goodstein, Laurie. "Study Finds One in Six Follows No Religion." The New York Times (18 Dec. 2012): n. pag. Print.

Harrison, Paul. "Naturalistic, Scientific Pantheism." Is Your Spiritual Home Right Here on Earth? World Panthesism, 18 Dec. 2012. Web. 25 Apr. 2016. <http://www.pantheism.net/beliefs.htm>.

Hassaballa, Hesham A. *The Beliefnet Guide to Islam.* New York: Doubleday-Random House, 2006.

Jacoby, Susan. "The Blessings of Atheism." The New York Times 6 Jan. 2013: n. pag. Print.

Johnson, Luke Timothy. *Great World Religions--Christianity (The Great Courses)*. Chantilly: The Teaching Company, 2003.

Mann, Charles C. "The Birth of Religion." National Geographic June 2012: n. pag. Web.

Meacham, Jon. "The Decline and Fall of Christian America." Newsweek 13 Apr. 2009: n. pag. Web.

—. "Rethinking Heaven." Time Magazine 16 Apr. 2012: n. pag. Print.

Pagels, Elaine. *Reading Judas*. New York: Penguin, 2007.

Metzger, Bruce M. *The Holy Bible/New Revised Standard Version*. Oxford: Oxford University Press, 1989.

Muesse, Mark W. *Great World Religions—Hindusism* (The Great Courses.) Chantilly: The Teaching Company, 2003.

Presbyterian Church. *The Book of Confessions. Louisville: The Office of the General Assembly*, 2004.

Rodwell, J.M. *The Koran* (English Translation.) New York: Random House, 1993.

Spong, John Shelby. *Eternal Life: A New Vision*. New York: Harper Collins, 2009.

—. Unbelievable: Why Neither Ancient Creeds nor the Reformation Can Produce a Living Faith Today. HarperOne, 2018.

—. *Jesus for the Non-Religious*. New York: Harper Collins, 2008.

—. *A New Christianity for a New World*. New York: Harper Collins, 2001.

—. *Re-Claiming the Bible for a Non-Religious World*. New York: Harper Collins, 2011.

Sullivan, Andrew. "Forget the Church, Follow Jesus." Newsweek 9 Apr. 2012: n. pag. Print.

Swearer, Donald K. *Becoming the Buddha*. Princeton: Princeton University Press, 2004.

SCIENCE

Abrams, Nancy Ellen, and Joel Primack. *The New Universe and the Human Future: How a Shared Cosmology Could Transform the World*. New Haven: Yale UP, 2011. Print.

Bronowski, Jacob. *The Ascent of Man*. New York: Random House, 2011.

Capra, Fritjof. *The Tao of Physics*. Boston: Shambhala Publications, 2010.

Carroll, Sean. *The Particle at the End of the Universe*. London: Penguin, 2012.

Cosmos: A Spacetime Odyssey. Dir. Ann Druyan and Neil DeGrasse Tyson. Twentieth Century Fox, 2014. DVD.

Cosmos: A Personal Voyage. Dir. Carl Sagan. Carl Sagan Productions, 1980. DVD.

deGrasse Tyson, Neil. *The Inexplicable Universe*. (The Great courses.) Chantilly: The Teaching Company, 2012.

—. *My Favorite Universe*. (The Great courses.) Chantilly: The Teaching Company, 2003.

deGrasse Tyson, Neil and Donald Goldsmith. *Origins*. New York: W. W. Norton, 2004.

The Dalai Lama. *The Universe in a Single Atom*. New York: Random House, 2005.

Darwin, Charles. *The Descent of Man, and Selection in Relation to Sex*. Princeton, NJ: Princeton UP, 1981. Print.

—. *The Origin of the Species by Means of Natural Selection (Barnes & Noble Classics)*. New York: Barnes & Noble, 2004.

Dawkins, Richard. *The God Delusion*. New York: Houghton Mifflin, 2006.

Dowd, Michael. T*hank God for Evolution*. New York: Penguin, 2009.

Filippenko, Alex. *Understanding the Universe (The Great Courses.)* Chantilly: The Teaching Company, 2007.

Grassie, William. *Politics by Other Means: Science and Religion in the 21st Century.* Bryn Mawr, PA: Metanexus Institute, 2010. Print.

Greene, Brian. The Elegant Universe. New York: W. W. Norton, 2003.

—. "The Mystery of the Multiverse." Newsweek 28 May 2012: n. pag. Print.

Hawking, Stephen. *A Brief History of Time.* New York: Bantam Books/Random House, 2005.

—. *The Grand Design.* New York: Bantam Books/Random House, 2012.

—. *The Universe in a Nutshell.* New York: Bantam Books/Random House, 2001.

Holt, Jim. *Why Does The World Exist?.* New York: W. W. Norton, 2012.

Journey of the Universe. Dir. Mary Evelyn Tucker. Yale University/Dolby Laboratories, 2011. DVD.

Meyer, David. *Experiencing Hubble (The Great Courses.)* Chantilly: The Teaching Company, 2011.

The New Universe: Here, Now, and Beyond. Washington, DC: National Geographic Society., 2010. Print.

Pinker, Stephen. *Enlightenment Now.* New York: Penguin Random House, 2018.

Polkinghorne, John. *Quantum Physics and Theology.* New Haven: Yale University Press, 2007.

—. *The Quantum World.* Princeton: Princeton University Press, 1992.

Primack, Joel. *The View from the Center of the Universe.* New York: Riverhead Books, 2006.

Sagan, Carl. *Cosmos.* New York: Ballantine Books, 1980.

—. *The Demon-Haunted World.* New York: Ballantine Books, 1996.

—. *Pale Blue Dot.* New York: Ballantine Books, 1994.

Swimme, Brian. *The Universe Story.* San Francisco: Harper Collins, 1992.

Wilson, David Sloan. *Evolution for Everyone.* New York: Random House, 2007.

Wolfson, Richard. *Einstein's Relativity and Quantum Revolution (The GreatCourses.)* Chantilly: The Teaching Company, 2000.

Wright, Robert. *The Evolution of God.* New York: Little, Brown and Company, 2009.

SOCIETY

Carter, Jimmy. *Our Endangered Values.* New York: Simon and Schuster, 2000.

—. *A Call To Action.* New York: Simon and Schuster, 2014.

Diamond, Jared. *Guns, Germs and Steel.* New York: W.W. Norton & Co.,1997.

The Economist: Pocket World in Figures, 2015 Edition. London: Profile Books, 2014.

The Economist: The World In 2014. London: The Economist Newspaper Ltd., 2014.

Friedman, Thomas L. *The World Is Flat.* Farrar, Straus and Giroux, 2005.

—. *Hot, Flat and Crowded.* New York: Farrar, Straus and Giroux, 2008.

—. *That Used To Be Us.* New York: Farrar, Straus and Giroux, 2011.

Janssen, Sarah. *The World Almanac and Book of Facts.* New York: World Almanac Books, 2014.

Jaspers, Karl. *The Origin and Goal of History.* English Translation, New Haven: Yale University Press, 1968.

Kagan, Neil. *Concise History of the World.* Washington: National Geographic, 2006.

Naisbitt, John. *Megatrends 2000.* New York: William Morrow & Co., 1990.

—. *Megatrends.* New York: Warner Books, 1982.

Podesta, John. "Relentlessly High Youth Unemployment Is a Global Time Bomb | John Podesta." *The Guardian.* Guardian News and Media, 09 July 2013. Web. 24 Apr. 2016.

Scaruffi, Peter. "The Worst Genocides of the 20th and 21st Century." *The Worst Genocides of the 20th and 21st Century.* Web. 24 Apr. 2016.

Shah, Anup. "Poverty Facts and Stats." N.p., Jan. 2014. Web. <http://www.globalissues.org/article/26/poverty-facts-and-stats>.

Shermer, Michael. *The Moral Arc.* New York: Henry Holt and Company, 2015.

Stanley, Arthur Penrhyn, *The Life and Correspondence of Thomas Arnold.* London: B. Fellowes, 1845.

Young, Andrew. *An Easy Burden.* New York: HarperCollins Publishers, 1996.

Image Credits:

P. 23 *Martin Luther King Jr. Mural at the National Historic Site in Atlanta, GA*
Credit: Forty3Zero / Shutterstock.com

P. 30 *Bust of Karl Jaspers*, 1983, Oldenburg. Credit: Christa Baumgärtel.

P. 42 *HH 901/902 .*Photo courtesy of NASA, ESA, and M. Livio and the Hubble 20th Anniversary Team (STScI)

P. 44 *The Creation of Adam* by Michaelangelo. Sistine Chapel. Public Domain

P. 49 *Refugees along the Slovenia/Austria border.* October 2015. Credit: Janossy Gergley/shutterstock.com.

P. 57 *Leh, India—June 29, 2015: Unidentified poor Indian beggar family on street in Ladakh. Children of the early ages are often brought to the begging profession.* Credit: Oleg D./Shutterstock.com

P. 65 *New York Stock Exchange.* Credit: Javen/shutterstock.com.

P. 68 *Bashik Frontline, Kurdistan, Iraq—Two unidentified kurdish (peshmerga) fighters in back of truck at Bashik (bashik) base 25km from ISIS controlled Mosul.* Credit: Owen Holdaway

P. 70 *Muslim veiled women in the heart of downtown Istanbul, Turkey.*
Credit: meunierd / Shutterstock.com

P. 72 *Martin Luther King, JR. Washington, D.C., 1963.* Credit: National Archives and Records Administration.

P. 76 *People praying in a Caodai temple in Vietnam.* Credit: Maurizio Biso/shutterstock.com

P. 84 *Ruins of Nagasaki, Japan.* Credit: Everett Historical

P. 90 *Isaac Newton.* Godfrey Kneller, 1689. Credit: Public Domain.

P. 94 *The Battle of Stalingrad.* Credit: Public Domain.

P. 102 *The Last Supper of Christ* by Guillaume Herreyns (1743-1827). Credit: Renata Sedmakova/Shutterstock.com

P. 106 *Charles Darwin, Age 45, 1854.* Credit: Everett Historical

P. 107 *The Vitruvian Man, Leonardo Da Vinci.* Credit: Imagineerix

P. 112 *International Space Station with astronauts over planet Earth.* Credit: Vadim Sadovski

P. 119 *The Hubble Space Telescope in orbit.* Credit: Marcel Clemens

P. 120 *Famed Astronomer Carl Sagan.* Public Domain

P. 121 *The 14th Dalai Lama.* Credit: Nadezda Murmakova

P. 127 *Large Hadron Collider at CERN Laboratory.* Credit: Thomas Jurkowski

P. 131 *Neurons in the Brain.* Credit: Crevis

P. 134 *Our Milky Way Galaxy. Credit: Standret*

P. 149 *The Devil. Credit: Vera Petruk*

P. 151 *Human bones in the crematorium of Buchenwald concentration camp after liberation. U.S. 3rd Army arrived at the camp near Weimer, Germany, on April 11, 1945.*
Credit: Everett Historical

P. 152 *New York, USA—Sep 27, 2015: President of Ukraine Poroshenko Petro delivers his speech at the UN Sustainable Development Summit in New York*
Credit: Drop of Light/Shutterstock.com

P. 154 *Andrew J. Young at the LBJ Presidential Library, 2013.* Public Domain

P. 182 *Smartphone users on Hong Kong subway.* Credit: eXpose/shutterstock.com

BLAINE KELLEY

VISIONARY DEVELOPER, HUMANITARIAN LEADER, PROGRESSIVE ADVOCATE

Blaine Kelley, Jr. is a resident of Atlanta, Georgia and a native of Charlotte, North Carolina. Mr. Kelley graduated from Davidson College in 1951 with a Bachelor of Science Degree in Economics. Kelley then served on active duty as a Lieutenant in the United States Army Artillery, and later as a Captain in the North Carolina National Guard.

Moving to Atlanta in 1957, Mr. Kelley married the former Sylvia Sanders, a native of Atlanta. A lifetime advocate of leadership and equality for women, Sylvia Kelley also has been a recognized leader and pioneer, both locally and nationally, for her work in educational and religious causes.

While Mr. Kelley's professional career began with management roles in manufacturing and construction, most of his career has been devoted to real estate development. In 1968 Kelley founded The Landmarks Group, serving this real estate firm as its Chief Executive Officer for twenty five years. Over this period of time, Landmarks developed commercial and residential properties collectively valued at several billion dollars. Recognized as a pace-setting leader in its profession, Landmarks was unique in its early inclusion of women and minorities as corporate officers. Mr. Kelley is now retired from business-related activities, devoting his full time efforts to community, humanitarian and educational causes.

While known locally and nationally as a pioneer property developer, Kelley has also been a leader in several major professional organizations. He has served as a Trustee of The Urban Land Institute, considered the leading real estate leadership organization worldwide. He served as a Trustee of The National Realty Committee, the industry's liaison with the U.S. Congress, as a Trustee of the building industry's historic National Building Museum and as a Trustee of the National Alliance of Business, a U.S. Congress established jobs and training organization. All of these are national professional organizations headquartered in Washington.

Kelley served for several terms as a Director and Vice Chairman of The Atlanta Chamber of Commerce. During his Atlanta Chamber service, Kelley participated in several overseas trade missions on behalf of new business for Atlanta. His Chamber related services also included Director of The Georgia Chamber of Commerce and Central Atlanta Progress. Kelley also served as Director of two major banks, the latter being the predecessor to the Bank of America of today.

Throughout his professional career, Mr. Kelley has been equally active in non-professional causes. Kelley served for three terms as a member of The Board of Trustees of Davidson College in North Carolina, where he also was the leader in the founding of a center for international studies and President of the Davidson College National Alumni.

Blaine Kelley holds the distinction of having served in an official capacity at five other major universities. These associations include: the Advisory Boards of Emory University's Goizueta School of Business, Duke University's Sanford School of International Studies, Vanderbilt University's Divinity School, Georgia State University's Andrew Young School of Policy Studies, and Harvard University's Divinity School and the Divinity School's Council on Public Policy.

In other areas, Mr. Kelley served as a member of the Board of Councilors of the Carter Center, the official Presidential center supporting international causes, as a Director of The Woodruff Arts Center, Atlanta's leading arts organization, and as Trustee of The Southern Center for International Studies, promoting the education and advancement of international relations.

In earlier years, Kelley also served as a member of the Board of Advisors of the Martin Luther King Jr. Center, a national organization devoted to the advancement of civil and human rights.

Another area of special interest for Blaine Kelley has been his membership in the 500 member Rotary Club of Atlanta. There he recently served as the primary leader of the club's "Project for Safe Water and Health in Kenya". Under Kelley's leadership, this extraordinary project brought together a partnering with several international organizations, each collaborating to offer life-saving water and health related products and services in two areas of The Republic of Kenya. The project has been recognized as one of the world's largest and most effective projects of its kind, bringing safe water and health items to over 400,000 persons, resulting in the saving of thousands of lives, especially those of young children and winning selection by a European-based world organization as the "World Humanitarian Project of the Year."

Another central activity of Blaine Kelley has been his work in religious related causes, much of this in partnership with his wife Sylvia. Mr. Kelley has served as a Ruling Elder of two Atlanta Presbyterian Churches, later serving on the Advisory Boards of Vanderbilt Divinity School and Harvard Divinity School.

During the 1980's Blaine Kelley was honored by the *Atlanta Business Chronicle* as Atlanta's "Executive of the Year", by the American Jewish Committee's Human Relations Institute as "Atlanta Humanitarian of the Year," by *Georgia Trend Magazine* as one of the "One Hundred Most Influential People in Georgia," and by The Associated General Contractors of Georgia as their "Honoree of the Year."

Additionally, Kelley has received the "Alumni Service Medal" of Davidson College and was selected by *Business To Business Magazine* for their Hall of Fame as one of the "Six Legends of Atlanta Real Estate".

The Rotary Club of Atlanta has recognized Kelley with its "Service Above Self Award". Kelley and his wife have been inducted into "The Collegium of Scholars and Sponsors" of the Martin Luther King Jr. International Chapel at Morehouse College in Atlanta. Additionally, Kelley was honored in 2013 by *The Atlanta Business Chronicle* as "Atlanta's Visionary of the Year," and in 2014, Kelley was honored by the local chapter of The Urban Land Institute with its "Community Leadership Achievement Award."

As a culmination of his life of multi-faceted leadership and diversity of experiences, Mr. Kelley has turned to a new role as an author. His current book is *The Global Imperative*, a treatise urging the rethinking and repurposing of religion and science as the most effective path to a new age of world peace, justice and equality.

Blaine Kelley